BIG SCIENCE ᶠᵒᴿ LITTLE PEOPLE

BIG SCIENCE
FOR LITTLE PEOPLE

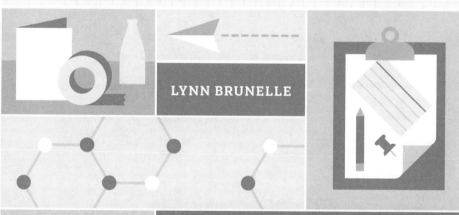

LYNN BRUNELLE

52 Activities to Help You & Your Child Discover the Wonders of Science

ROOST BOOKS
BOULDER
2016

ROOST BOOKS
An imprint of Shambhala Publications, Inc.
4720 Walnut Street
Boulder, Colorado 80301
roostbooks.com

9 8 7 6 5 4 3

Printed in the United States of America

♾ This edition is printed on acid-free paper that meets the American National Standards Institute z39.48 Standard.
♻ Shambhala Publications makes every effort to print on recycled paper. For more information please visit www.shambhala.com.

Roost Books is distributed worldwide by Penguin Random House, Inc., and its subsidiaries.

Designed by Daniel Urban-Brown

LIBRARY OF CONGRESS CATALOGING-IN-PUBLICATION DATA

Names: Brunelle, Lynn, author.
Title: Big science for little people: 52 activities to help you and your child discover the wonders of science/Lynn Brunelle.
Description: First edition. | Boulder, Colorado: Roost Books, [2016] | Series: An official geek mama guide
Identifiers: LCCN 2015045944 | ISBN 9781611803501 (pbk.: alk. paper)
Subjects: LCSH: Science—Miscellanea—Juvenile literature. | Science—Experiments—Juvenile literature.
Classification: LCC Q164 .B854 2016 | DDC 502—dc23 LC record available at http://lccn.loc.gov/2015045944

For the loves of my life:
Kai and Leo,
my intrepid explorer-scientists,
and Keith, my favorite lab partner
who puts the fizz and sparkle
in everything!

CONTENTS

ON THE PLAYGROUND: LOUD, MESSY, AND FREE-RANGE EXPERIMENTS

69

INTRODUCTION

I'LL LET YOU IN ON A SECRET. I am a geek. (Maybe it's not a *huge* secret.) But more than that, I love being a geek! To me, being a geek means having an unabashed enthusiasm for learning, wondering, and questioning. It means allowing myself to be passionate about the hows, the whys, and the whats. I am drawn to anything that has the slightest whiff of wonder. There's a LOT of wonder in science!

As a mom with a deep passion for science, I know that when the kids' imaginations get hooked they become curious, and then they start asking the big questions: What is that? How does it work? Why does that happen? As a Geek Mama, I know that when *that* happens, a scientist is born.

Science is a glorious way to play and learn. Children are natural born scientists already. Every day their playing and exploration and discovery deepen their understanding of the world around them. They approach the world with great curiosity and imagination. Hands-on explorations help them to separate reality from fantasy. They ask questions, they make observations, they gather information, and they compare data and identify patterns without the help of any textbook. What we do as parents is to offer them the world of the science lab. We get them to describe and discuss their adventures and observations and to form explanations. This is science!

As a parent, there is nothing more fun than watching your kids delight in something with wide-eyed amazement. And, as a kid, there is nothing more fun than making something bubble, fizz, explode, or do something surprising. This book is a celebration of this sort of joy, with a very simple premise: when you invite children to explore ordinary materials through a scientific lens, they get an up-close look at the astonishing and mind-expanding way that things work. And, it's fun!

Between the covers of this book lies a world of potions, fizzy mixtures, mini explosions, and really fun and accessible science activities. Here are the building blocks for a budding interest in physics, biology, chemistry, and (gasp!) math. This book is just as much fun for you as it is for your kids. Tap into that inner geek of yours—it's there—and let yourself wonder right along with the kids. Delight in Exploding Sidewalk Chalk and Pop Stick Explosions. Be mystified by a Mobius strip. Make paper buzz and balloons scream. Turn the garden hose into a tuba and a straw into an oboe.

All of the experiments here are more than just fun, too. They're about real science—exploring, observing, and experimenting. Kids will dabble in the physical forces that keep our universe in motion. The activities presented here are easy, and the concepts are accessible. There are suggestions for taking the activity further, and sidebars and boxes explain the technical stuff. Whether or not the kids want to get technical, at least *you'll* know why bubbles aren't square, how airplanes fly, and why baking soda and vinegar explode. Once you know, you'll want to share. And so will they.

I've divided the activities into two major sections: those you can do inside—at the kitchen counter, in the bathtub, or at the table; and those you might want to do outside—in the yard, at the playground, on the sidewalk, or anywhere you can make a mess. With these fifty-two, easy-to-do science activities, projects, explorations, and experiments, I give you the moon and the sun and the permission to wonder, to make a mess, to laugh loud, and to be amazed.

Science is SO COOL! You'll see.

CREATING A HOME SCIENCE LAB

....

It really doesn't take much to have a truly awesome functioning science lab in your house. In fact, you probably already have the materials for most of these projects in your kitchen, garage, or bathroom. And when I say "science lab," think more "portable science bin." Whether it's water experiments in the bathtub, kitchen counter potions, or backyard explosions, this is a lab that little hands can easily dig out whenever the urge to wonder and explore comes upon them.

KITCHEN COLLECTIONS

Your kitchen is probably already stocked with the items on this list. Commonly used for various activities like baking, pickling, cleaning, and cooking, these things are also essential materials for your lab. You don't have to keep them in your bin, but be

+ Baking soda
+ Corn starch
+ Food coloring
+ Vinegar
+ Dish soap
+ Raisins
+ Vegetable oil
+ Zipper-lock plastic bags
+ Eggs
+ Milk
+ Bread
+ Marshmallows
+ Plastic drinking straws
+ Salt
+ Alum
+ Borax

sure to have them on hand for fizzing, foaming, forming, and fabricating.

RANDOM RECYCLABLES

You're going to want to start saving some things that you previously thought of as trash. Here's a handy guide for saving those items that would ordinarily be recycled and adding them to your fully functioning lab. In addition to recyclables, a number of items you probably already keep in your art bin, office, garage, or junk drawer can also come in handy.

+ Scissors
+ Ruler
+ Stapler
+ Tape
+ Glue
+ Index cards
+ Paper
+ Paper clips
+ Rubber bands
+ Pushpins
+ Popsicle sticks

+ String
+ Balloons
+ Coins
+ Washers
+ Hex nuts
+ Hammer
+ Nails
+ Paper towel
  tubes
+ Toilet paper
  tubes

+ Newspapers
+ Magazines
+ Paper cups
+ Paper plates
+ Yogurt
  containers
+ Plastic pop
  bottles
+ Take-out
  containers

CREATING A SPACE FOR SCIENCE

Everyday life is brimming with science experiences and opportunities to explore and learn. Having a bin with materials and supplies is a great start to giving kids an inviting atmosphere for experimenting. Offering the opportunities to ask and wonder is one thing, but having a place where you can make a mess and revel in it is another. We have a few places where it's OK to just play and ooze and mess and goo. The kitchen counter is one place; the bathtub is another; outside is always a favorite. My motto for my kids is "Make a mess! That's where you learn stuff. But remember, the experiment isn't over until the lab is cleaned up." As a parent you can direct them to scenarios where they can be successful on all counts!

THINGS I HAVE LEARNED ALONG THE WAY

I can have the best intentions. I can be prepared for almost anything, and *inevitably* things take a turn. Like the time Kai and Leo and I decided to make homemade bubble goo and blow bubbles. We never got to the bubble part. Instead they were fascinated with the slickness of the goo. It evolved into a homemade slip and slide in the front lawn with the hose, a huge trash bag, and some biodegradable dish soap—and the screams and hollers of pure joy. If I had kept them on task I am sure they would have had fun, but what they veered into was fantastic! And it was truly their journey.

Over and over I learn that tagging along on their creative direction can be really wonderful. Of course as a grown-up I can't let go when it comes to keeping them safe, and it's not easy for me to abandon my plans—they always seem so great to *me*—but when I do let go and allow my kids to drive, it allows them to have some control and passion and really get into what they find fascinating. In the long run this is such a great gift for them. Doors are opened. The world is delightful! Passions are planted.

So I offer you my hard-earned wisdom. These experiments are designed to deliver an engaging science experience. At the end of every activity I have written a few extra suggestions for extending the play. These ideas are based on my own experiences with my kids. Some of them are areas where they veered. Use them as springboards. Jump in, have fun, and see where they take you!

AT HOME

KITCHEN CONCOCTIONS, BATHTUB TESTS, AND CREATIVE EXPLORATIONS

Science is literally everywhere! This means that a lot of science can be explored right at your fingertips and under your nose in your own house! The kitchen, for example, is a wonderful laboratory: layer some liquids, dissolve an eggshell, make raisins dance, grow mold, and hang a spoon on your nose. While you're at it you'll be dabbling in the magic of chemistry, physics, and biology.

Take the action to the bathroom, and the tub is a wonderful world to explore gravity, buoyancy, and air pressure. Make a tornado. Make water float on air. Float things! Poke holes in a bag and don't even spill a drop! Sink things!

And if that isn't enough to float *your* boat, head to the office. The things you can do with ordinary office supplies will knock your socks off. Make a paper snake dance, walk through an index card, and move a paper clip using only your voice.

These projects are self-contained and are easy to do at mealtime, bathtime, and quiet playtime. They will stir the imagination and spark a curiosity in the world we live in.

(1)

LAYERING LIQUIDS

Stacking things is always fun. Blocks, yogurt cups, books, and even firewood are solid and can balance on top of each other. That's not surprising. But you can actually stack some liquids, too.

Note: I didn't put hard and fast measurements here because it's all about layering in the bottle you have. Eyeball it. About an inch or two of each liquid is enough.

STUFF YOU NEED

One clear 2-liter plastic pop bottle or other clear glass or plastic bottle
Scissors
Corn syrup
Food coloring
Water
Vegetable oil
Rubbing alcohol
Metal spoon

THINGS YOU DO

❶ Cut the top off the plastic pop bottle so you have a large open cylinder. This will be your liquid-layer holder. Pour in about 2 inches of corn syrup.

❷ Now start layering the liquids. First, make a prediction about what will happen when you add the water. Add a few drops of food coloring to the water, then pour in the water and see what happens. [CONT.→]

Second, predict what will happen when you add the vegetable oil. Pour in about 2 inches of oil.

Finally, predict what will happen when you add the rubbing alcohol. Add a few drops of another color of food coloring to the rubbing alcohol and gently pour it in, using the spoon to softly layer the alcohol on top of the liquids.

❸ What do you see? Why don't the layers mix?

 ## WHAT'S GOING ON?

This experiment is all about density. Technically speaking, *density* is how much mass is in a given volume. That's technically speaking. How I like to describe it is like an elevator. If you've ever been in an elevator when there is a ton of people in it, you are squished into a tiny space and there is not much room between people. You can't move around much. That elevator is pretty dense. Now take the same elevator but you are in it on your own with maybe one friend. You guys can dance, spin, whip around, and bounce off the walls if you want to. (But you don't want to do *that*!) There is a lot of space in that elevator and not so many people. This elevator is not so dense. It's the same with stuff like liquids, solids, and gases. A dense material is jam-packed with molecules. A less dense material has fewer molecules filling up the same space.

In this experiment, the corn syrup is the most dense. It's the crammed elevator. It sinks to the bottom. The water floats on top of the syrup. It is a less crammed elevator. The oil floats on the water, and the alcohol floats on top of everything because it is the least dense of all the liquids.

Try tossing solid things into your layered liquids. Solids have different densities as well, and it's fun to see where things land. Try objects such as a penny, a cranberry, a toy dinosaur, a marble, a paper clip, and so on. Make a chart and predict on which layers things will finally rest.

2

POP BOTTLE LAVA LAMP

Plastic soda bottles are the best little minilabs! They are water-proof, airtight, and see-through, on top of being relatively unbreak-able and portable. Some of my favorite experiments can be done with this remarkable recyclable. This pop bottle lava lamp is no exception. You will make layered liquids, start a chemical reaction, and see a really cool display of density play out in your bottle.

STUFF YOU NEED

Water

One 1-liter plastic pop bottle

Funnel

Vegetable oil

Food coloring

Alka-Seltzer–type fizz-ing antacid tablets

THINGS YOU DO

❶ Pour about 2 inches of water in the bottom of the bottle. Using the funnel, fill the rest of the bottle with oil. Fill to about 1 inch from the top.

❷ Make a prediction about what will happen when you put food coloring in and then add about 6 drops of your favorite color. The food coloring is water based, so it just drops right through the oil and sits on top of the water. Don't worry! It won't stay there for long.

❸ Open up a packet of the fizzy tablets and break one in half. Again, predict what will happen before you drop in the tablet. Drop in the tablet and watch.

Caution: DO NOT PUT THE CAP ON! If you do, the gas will build up and you could have a dangerous explosion.

WHAT'S GOING ON?

The water is more dense than the oil, so it sits at the bottom of your bottle. The tablet doesn't react with the oil, so it won't fizz. But since the tablet is more dense than the oil, it sinks. As soon as the tablet hits the water, *kaBLAM!* It reacts with the water and produces carbon dioxide gas. This gas is *way* less dense than the water and the oil, and it rises fast. It rises so fast that it takes the water with it. When the gas escapes from the surface, the water sinks back down. That's why you get the rolling up and down motion of the bubbles. And since the food coloring is only soluble in the water, the bubbles are colored.

❓ TAKE IT FURTHER

+ When the bubbles stop, you can add more fizzing tablets. This project can keep going and going and going as long as you add the tablets.
+ Since we cannot see the gas, how do we know what is happening? Put a balloon over the mouth of the bottle after you add a tablet and watch it collect the gas that is escaping.
+ Turn out the lights and put a flashlight under the bottle to create a glowing lava lamp.

EXPLODING LUNCH BAG

How can you make lunch more fun? With an exploding lunch bag! This time-release baking soda lunch bag blaster is a fun challenge: can you zip the bag closed before the reaction happens? You will want to do this one in the sink or outside. If you take it outside, it's best to do it on a sidewalk or a driveway because the vinegar can kill grass.

STUFF YOU NEED

Tissue

2 heaping tablespoons baking soda

Zipper-lock sandwich bag

White vinegar

THINGS YOU DO

❶ Open up a tissue and place the baking soda in the middle. Fold the sides in and roll it like a burrito.

❷ Fill a sandwich bag ½ full of vinegar, and zip it mostly shut.

❸ Now, drop the burrito-bomb in the bag. Zip it shut immediately! Give the bag a little shake, and when it starts to expand, place it in the sink or on the ground and walk away, fast!

GEEK MAMA FUN FACT!

This experiment has a great added benefit. Mixing baking soda and vinegar is a great way to clean the sink! Once it explodes, give the sink a little swaggle with a sponge and you'll have a bacteria-free and sparkly sink!

 WHAT'S GOING ON?

When you mix an acid (like vinegar) and a base (like baking soda), they create a chemical reaction. They fizz and create a gas called carbon dioxide. The carbon dioxide builds up and pushes against the inside of the sandwich bag until something gives. *BLAM!* The bag pops.

 TAKE IT FURTHER

+ Try this experiment with different-sized plastic bags. What happens if you have more vinegar? What happens if you add more baking soda?
+ In the summertime, we add the element of sidewalk chalk to our plastic bag explosions. See page 87 to make Exploding Sidewalk Chalk!

MOLD GARDEN

Don't have a garden or a pea patch? No problem. You can grow a little mold garden right in your own kitchen in a little zipper-lock bag. It's amazing to me that living things are everywhere around us even if we can't see them. In this activity, you'll capture some of these invisible living things and grow them in numbers big enough so you can see them.

STUFF YOU NEED

Piece of bread
Spray bottle and water
Zipper-lock sandwich
bag

THINGS YOU DO

❶ Take a piece of bread and spray it lightly with the water sprayer. Place it on the counter of your kitchen or outside on a surface for 1 hour or so.

❷ Spray it one more time lightly and put the whole slice into the bag. Zip it shut, and let it sit in a warm place out of direct sunlight for a few days. Keep checking on it. What do you see?

Caution: Do not open the bag after the mold has grown. This will spread spores everywhere, and some may be bad for your health.

If you're lucky you have grown a few fuzzy spots of mold on your bread. But what is mold, anyway? Mold is a kind of fungus. The mushrooms on your pizza are a kind of fungus, too. That doesn't mean you'll be able to grow mushrooms on your bread, though. And even if you did, it wouldn't be a good idea to munch on them.

Unlike plants, molds don't grow from seeds. They grow from tiny spores. These spores are everywhere, but they are so tiny you can't see them. They float in the air, and when they land on something damp with nutrients, they set up a home and grow into mold.

Molds are not plants. Plants are green and have a chemical called chlorophyll in their leaves that helps them make energy and food from the sun. Molds can't do that. They have to find food, like bread or cheese. The spores land on food and then spit out chemicals that make the food start to rot. As the food rots, the mold thrives.

? | TAKE IT FURTHER

+ There are a ton of different kinds of mold in our world. The gray fuzzy kind lives on strawberries and the blue powdery kind grows on lemons. Get a bunch of plastic zipper-lock sandwich bags and see what grows on some of the foods in your house. What grows on oranges, blueberries, and broccoli?
+ Some foods spoil faster than others. Why? Try growing mold on a food that has a lot of preservatives, like a store-bought

packaged cookie or cupcake. Try putting homemade cookies in another. Which one rots faster?

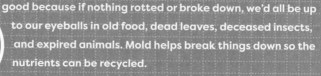

5

DANCING RAISINS

This experiment is a popular one in our house. Making the potion fizz is mesmerizing, of course, but tossing things adds a new dimension. Even better than all that is seeing the raisins start to dance. They go up and down on a kind of Ferris wheel of bubbles. No vinegar? No problem! If you don't have vinegar and baking soda on hand, a glass of clear seltzer or pop will work.

STUFF YOU NEED

1 clear drinking glass or empty, clear bottle
½ cup water
½ cup white vinegar
1 tablespoon baking soda
Handful of raisins

THINGS YOU DO

❶ Add the water to the glass. Then add the vinegar.
❷ Toss in the baking soda. *Ooh* and *ahh* over the fizz. Gently toss a handful of raisins into the fizzy soup. Wait a bit and watch.

❗ WHAT'S GOING ON?

When you mix vinegar (an acid) and baking soda (a base) together, they react and form carbon dioxide (a gas). Gas is much less dense than liquid, and so it floats up. Raisins are more dense than water, so they sink. But raisins are also wrinkly. The bubbles of carbon dioxide get trapped in those wrinkles and haul the raisins up as the bubbles go up. When the raisins get to the top, the gas escapes and the raisins drop back down.

❓ TAKE IT FURTHER

+ Does it only work on raisins? Hmmmm. Check out some options. Craisins, anyone? Peanuts? Candy conversation hearts? M&Ms? Grapes? What works? What doesn't?
+ How long can they dance? Time those raisins!
+ Gas is invisible when it comes out of the liquid. How do we know it's there? After you have added all the goodies from Things You Do, seal a balloon over the top and watch it inflate.

6

DISAPPEARING EGGSHELL

Everyone can crack an egg, but this experiment lets you make the hard outer shell of an egg disappear. A common kitchen ingredient will simply dissolve it, leaving behind a watery bag of egg! It takes a bit of time, but the result is worth it.

STUFF YOU NEED

1 egg

Clear glass jar with lid
(An old jam jar works well.)

White vinegar

THINGS YOU DO

❶ Gently place the whole egg in the jar. Pour the vinegar over the egg until it is covered completely. Watch. You should start seeing bubbles right away.

❷ After 2 days, change out the vinegar. After 4 days remove the egg and gently wash it under cool water. What do you notice?

Caution: Do not eat the egg!

 WHAT'S GOING ON?

Eggshells are made up of calcium, the same stuff that makes up your bones. When calcium comes into contact with an acid, it dissolves. Vinegar is an acid, and when it comes into contact with the calcium carbonate of the eggshell, it changes the structure of that chemical and dissolves it. What you are left with is the tough membrane that surrounds the egg yolk and egg white.

+ Try soaking your shell-less egg in a solution of 1 cup of water to 1 cup of corn syrup. Let it soak for a few hours. What happens?
+ What happens when you try this with a hard-boiled egg? Try it and see!

⟨7⟩

BALANCING FORKS

Is dinner taking a little longer than expected to arrive at the table? Invite science along to pass the time. Here's a mystifying experiment that seems impossible at first. All you need are a couple of forks, a toothpick, and a glass.

STUFF YOU NEED

Drinking glass
2 matching forks
Toothpick

THINGS YOU DO

❶ Place the glass on the table.
❷ Wedge the 2 forks together by interweaving their tines. They should stick together when you pick them up.
❸ Find the midpoint of the forks by balancing them on your finger. Now place the toothpick through that point. Carefully place the toothpick on the rim of the glass so it balances.

 WHAT'S GOING ON?

Every object has a center of gravity—a place where the forces and mass are balanced from left to right, front to back, up and down, and all around. If you balance a ruler on one finger, the point where your finger is when the ruler is balanced is the center of gravity. We all have a center of gravity. It's why we don't fall down all day. So, the forks have a center of gravity, and that's where you put the toothpick. When you balance the toothpick on the glass, take a look at the forks—they will be below the toothpick. The center of gravity is directly below the point where the toothpick is balanced (called the *pivot point*).

 ## TAKE IT FURTHER

Add a little WOW factor to this experiment by lighting it up. You'll need a grown-up and a match. Since the center of gravity is already in place here, you can get rid of the rest of the toothpick. Take a match and light the toothpick on the inside of the glass. Make predictions about what will happen and watch. The toothpick burns right up to the edge of the glass and the forks stay balanced!

GEEK MAMA FUN FACT!

 When you put on a backpack filled with schoolbooks, your center of gravity shifts. The same happens when a woman gets pregnant, and that's why her balance is sometimes affected.

MAGIC MILK

Watching a puddle of milk on a plate sounds about as interesting as watching paint dry, but add a little color and a drop of soap, and you will be amazed! This is simply dazzling. You'll want to do it again and again.

STUFF YOU NEED

Milk
Plate
Food coloring
Cotton-topped swab
Dish soap

THINGS YOU DO

❶ Pour some milk on a plate. Put 4 dots of food coloring on the milk in the center of the plate. Leave a bit of space around each dot of color. (Bonus: Try doing this as a North, South, East, West kind of thing. It's a conversation starter.)

❷ Dab the swab in the dish soap. Very carefully touch the dots of food coloring in the milk with the soapy swab. ShaZAM! The colors burst and mix and flow!

 WHAT'S GOING ON?

The secret of the gushing, roiling colors is in the chemistry of that tiny drop of soap. Soap molecules have two ends: a water-loving side (called *polar*) and a water-fearing side (called *nonpolar*). It turns out that milk is mostly water. It also contains vitamins, minerals, proteins, and tiny droplets of fat floating around.

When the soap is introduced to the milk, the soap weakens the chemical bonds that hold the proteins and fats in the solution. The soap's water-loving end dissolves in water, and its water-fearing end attaches to fat globules in the milk. This is when the fun begins. The molecules of fat bend, roll, twist, and contort in all directions as the soap molecules race around to join up with the fat molecules. During this fat molecule dance, the food coloring molecules are bumped and shoved everywhere, providing an easy way to observe all the invisible activity. As the soap becomes evenly mixed with the milk, the action slows down and eventually stops.

 TAKE IT FURTHER

+ Try using different color combinations.
+ Try using different kinds of milk. Does whole milk work better than skim milk?

(9)

RAINSTORM TO GO

Can't get to the rainforest? Bring the rainforest to you! There's no such thing as a dry spell when you have a clear plastic take-out container. Add some soil, plants, and water, and voilà! You have a beautiful, closed-system environment and a way to watch it rain inside. This project is a great way to recycle take-out containers, but a terrarium can be built in any clear glass or plastic container that can be closed.

STUFF YOU NEED

Small pebbles

Clean, clear plastic take-out container

Activated charcoal (You can get this at the fish tank supply area of a pet store.)

Potting soil

Small plants

Water and sprayer

Small toys, shells, or rocks for decoration

THINGS YOU DO

❶ Place a thin layer of small pebbles in the bottom of the container for drainage. Sprinkle a layer of activated charcoal over the stones (this helps keep things cleaner and less stinky). Next, put a layer of potting soil over the whole thing. Carefully plant your plants in the soil.

❷ Spray water inside the terrarium—about 10 squirts. Now you can decorate with small toys or shells or rocks. Close the lid and place the terrarium in a bright area—not in direct sunlight, though, as the plants could burn.

❸ Watch your environment. Notice the changes that happen. [CONT.→]

Water will condense on the top and rain down. It's the water cycle! You might have to add water every once in a while, but for the most part, the terrarium world should be independent—it will rain inside on its own!

 WHAT'S GOING ON?

Your take-out container environment can go for weeks or months without needing water. It's a closed system, which means the water doesn't escape, it just goes around and around. It's the water cycle. As air inside the terrarium heats up, water will be pulled up from the soil to the top of the container. That's called *evaporation*. It forms a mist on the roof of the container. The water molecules stick together. That's called *condensation*. Then, finally, the weight of the water is too much and it falls back down as rain. That's *precipitation*. Evaporation, condensation, and precipitation form the water cycle.

 TAKE IT FURTHER

Try making an ecosystem in a zipper-lock baggie. Make more than one and see how different plants fare. Or try putting versions of the same ecosystem in different places where exposure to light changes. Is there a difference?

SPOON ON YOUR NOSE

You're in the restaurant and you've ordered your dinner, but time passes and everyone starts to get fidgety. The solution? Hang a spoon from your nose, of course.

STUFF YOU NEED

Metal spoon
Breath (preferably fresh)
Nose

THINGS YOU DO

❶ Pick up the spoon. Take a deep breath and fog up the inside part of the spoon.

❷ Hold the fogged up spoon on your nose for a few seconds and then slowly move your hands away. If the spoon flops off, no problem. Try again. It's tricky. When the spoon does stick, pause for effect. Add a "TaDAH!"

 ## WHAT'S GOING ON?

There are a few science-y things going on to make this work: gravity, friction, and adhesion. *Gravity* is the force that pulls everything toward the Earth. It pulls you down, and it pulls on the spoon. *Friction* happens when two surfaces rub against each other. Because the spoon is curved inward and your nose is curved outward, they fit nicely together. The Earth's gravity pulls down on the spoon, but the friction between your nose and the spoon help to hold the spoon up. *Adhesion* is a big word that means "stickiness." When you breathe on the spoon you actually coat the bowl of the spoon with tiny droplets of water that come out of your lungs. (You've seen your breath before on cold days when you walk outside—a cloud comes out of your mouth. That cloud is actually tiny droplets of water that come from your lungs. It happens ALL the time—you can only see it on cold days.)

Normally water is kind of slippery—like when you go down a water slide. The water molecules slide past each other easily. But when you have only a very thin coating of water molecules, the molecules don't slide so fast and it can act like a kind of glue. (Have you ever licked your fingers in order to turn a page in a book or grab one of those plastic bags in the grocery store? It's the same thing. The thin layer of water molecules makes your fingers a bit stickier.)

Combine gravity, friction, and a thin coat of water molecules, and voilà! You get a spoon hanging from your nose.

 TAKE IT FURTHER

+ Who has the most elegant suspension?
+ Who has the most endurance? Time who can hang a spoon on his or her nose the longest.
+ Can you walk with a spoon on your nose? Try relay races.

THE UPS AND DOWNS OF HOT AND COLD

We all grew up hearing "heat rises." It's how weather works, how the Earth's tectonic plates move, and even how lakes and ponds mix things up. But the truth is, we have it backward! Heat doesn't rise, cold sinks (and pushes heat up). It's the same—but slightly different. This experiment is a great way to see the process in action.

STUFF YOU NEED

2 empty 1-liter pop bottles with the labels removed

Cold water

Hot water

Food coloring

Playing card

THINGS YOU DO

❶ Fill a bottle with cold water right up to the lip. Next, fill the other with hot water. (Hot water from the tap is fine; don't burn yourself or melt the bottle.) Put a few drops of red food coloring into the hot water.

❷ Place the playing card over the mouth of the cold water bottle. Press the card tightly to the open mouth and hold on to it while you flip the whole bottle over. Now place the card and the bottle over the open mouth of the hot water bottle. Line it up so the 2 bottle mouths are perfectly aligned.

❸ Gently slide out the card. (You may have a few drips at this point but try to keep the mouths of the bottles lined up.) You will have to hold the bottles in place. What happens?

In a *solid*, the molecules are packed together and have a strong bond. They have their own shape. A *liquid* has molecules with more space in between. These substances take the shape of their container. The molecules slide easily past one another. In a *gas*, there is a lot of space between the molecules. Gases take the shape of their container as well.

All fluids are made up of molecules. Think of a clear glass jar filled with small marbles. The marbles act like molecules and fill up the space inside the jar. There is space between the molecules. When you heat up a liquid, you can make the spaces between the molecules (or marbles) bigger. The molecules move faster and bounce around more, taking up more space. They take up more space because there are bigger spaces between the molecules. They become less dense. Cooler fluids are moving more slowly and take up less space and are more dense. They sink down and push the warmer stuff up.

 TAKE IT FURTHER

Try doing the experiment again, only this time put the colored hot water bottle on top. What happens?

12

MAKE WATER FLOAT ON AIR

In our house something is getting spilled all the time somewhere. But this experiment uses the power of the atmosphere to keep a full glass of water from spilling a drop—even when you turn it upside down!

STUFF YOU NEED

Glass

Water

Old playing card or index card

THINGS YOU DO

❶ Fill the glass to the very top with tap water. The water should come right up to the lip of the glass.

❷ Cover the glass with the card. Holding the card in place, turn the glass upside down. Don't let go! (Try this over the sink until you get the hang of it.)

❸ Carefully remove your hand, leaving the card in place. (Don't jiggle the card!) *Gasp!* It's so cool! Hold the water over your head if you dare! The card stays put and the water stays inside the glass even though it's upside down!

How do the playing card and the water defy gravity? It's all in the power of the atmosphere. Our atmosphere is a 1-mile blanket of air that covers the whole planet. All those air molecules have weight. They push on us all the time. They push down on us, in on us, and even up on us. The weight of the molecules pushing on us is atmospheric pressure. It's everywhere! We're used to it. When you placed the card over the glass, the pressure on both sides of the card was equal. Molecules were pushing down from the atmosphere and up from the glass. When you flipped it over, the pressure of the atmosphere kept pushing in on the card—it was in balance—holding the card in place. Bump the card, and *whoosh!* Gravity comes into play, and the water comes pouring out.

❓ TAKE IT FURTHER

+ Try putting soap around the lip of the glass, then repeat this experiment. Make sure you do this over the sink. What happens?
+ Try using a small piece of mesh screen (such as from a screen door) over the mouth of the glass instead of the card. What happens?

13

IMPOSSIBLY DRY PAPER

Looking for a rainy day activity? Fill the bathtub with water and grab a few plastic, see-through cups for some tub-time science fun. This experiment is simple yet mesmerizing. How can you keep paper dry even if it's underwater?

STUFF YOU NEED

Tissue (A napkin or paper towel will also work.)
Cup
Bathtub full of water

THINGS YOU DO

❶ Crumple the tissue. Push it deep inside the cup so it sticks and doesn't fall out when you turn the cup upside down.

❷ Fill a tub with water, and hop in.

❸ Hold the cup upside down—the mouth of the cup should meet the surface of the water first. Press the cup down into the water, keeping it straight. If you tilt it, water will flow in. Keeping it flat, push the whole cup underwater.

❹ Now lift up the cup—straight up. The paper stays dry!

WHAT'S GOING ON?

The cup has more than paper in it—it's also filled with air. When you press the mouth of the cup flat against the water's surface, the air stays trapped. That air pushes up against the paper and the bottom of the cup and pushes the water away. The tissue is surrounded by air and never gets near the water. If you tip the cup, though—*WHOOSH*! The paper will get soaked.

 TAKE IT FURTHER

+ What happens to the level of the water vs. air when you push the cup farther underwater?
+ Try this in a pool. Put on your goggles and watch what happens the deeper you go.

(14)

POUR AIR
UNDERWATER

You can't see air, but you can pour it. This is a concept that's hard for kids to grasp, because if you can't see it, how can you see it pour? Surround the air with water! Try pouring air underwater.

This is a fun tub-time experiment.

STUFF YOU NEED

Bathtub full of water
2 clear plastic cups (You can do it with plastic pop bottles, too.)

THINGS YOU DO

❶ Fill the tub with water and hop in.
❷ Take the 2 clear containers and look at them. Submerge one entirely and listen for the noise. That *flop* is the sound of water rushing in and kicking out the air. Turn the other container upside down and place it mouth down. Submerge it. Don't let the air out. You should have one container filled with water and the other filled with air.
❸ Move the water-filled container so its mouth is now pointing down. Position it directly above the container with the air. Slowly tip the container filled with air so that bubbles stream upward into the container filled with water. You are pouring air!
❹ Reverse the positions and see if you can pour it again.

Air is a gas. It is less dense than the water, which means it will always rise up. So the upside-down container held air. That air kept the water out. When you held and tilted the container of air under the container of water, the air rushed out and up. The surrounding water pushed into that cup. You caught the air in your other cup. It pushed the water out of the way. You were pouring air. Normally you pour things down and let gravity pull the liquid down. But since air wants to go up when it's underwater, you have to "pour" it up.

❓ TAKE IT FURTHER

Fill one cup with water and turn it so the mouth of the cup is pointing down. Pull the cup up and out of the water. If you don't tilt the cup or let the lip of the cup break the water's surface, what happens? Can you pour air into it like this?

TORNADO IN A BOTTLE

When I was small and saw *The Wizard of Oz* the first time, I was terrified by the tornado. Growing up in Maine, I had never seen such a thing. It was mysterious and scary to me. If I had only had this! Here's a portable tornado that not only teaches how tornados happen, but also is sparkly and fun to boot.

STUFF YOU NEED

2 empty 1-liter pop bottles with labels removed and caps off

Water

Drop of dish soap

Glitter

Metal washer (Find a washer that fits flush, or as close as possible, on the mouth of the bottle; the outer diameter of [CONT. →]

THINGS YOU DO

❶ Fill 1 bottle ¾ full of water. Add 1 drop of dish soap (this keeps things from getting nasty!) and 1 pinch of glitter (this makes it sparkly and fun!). Place the washer over the mouth of the bottle.

❷ Take the empty bottle and balance the mouth of it on top of the mouth of the other bottle; the washer needs to sit in place between the bottles. Use duct tape to tape the 2 bottles and washer in place. This [CONT.→]

the washer has to be the same diameter as the opening of the bottle.)
Duct tape

Make sure the washer fits so the outer edge is flush with the outer lip of the mouth of the bottle.

can be a bit tricky—you may need to wrap it a few times, or your tornado will leak. (Two weather disasters in one!)

❸ Watch your environment. Notice the changes that happen.

❗ WHAT'S GOING ON?

When you pour the water out of the top bottle, the water wants to get down into the bottom. Gravity is pulling on it. But that bottom bottle is not empty. It is full—of air! And that air is pushing like crazy to get from the bottom up to the top. The water and the air have to muscle it out and push past each other through a small opening.

When you give the bottle a swirl, you create a nice little superhighway. The air zooms right up the middle and the water swirls around the outside. It's called a *vortex*. A vortex is a type of motion that causes liquids and gases to travel in spirals around a center line. A vortex is created when a rotating liquid falls through an opening. Gravity is the force that pulls the liquid down into the hole and pushes the air back up through the hole. If you look carefully, you will be able to see the hole in the middle of the vortex that allows the air to come up inside the bottle. If you do not swirl the water and just allow it to flow out on its own, then the air and water have to essentially take turns passing through the mouth of the bottle, thus the *glug-glug* sound.

The tornado swirl in your bottle is just like a real tornado. Hot air builds up when the sun warms up the land. That hot air rises. Colder air, high up in the sky sinks. Add a little wind and *vaVOOM*, you can get a vortex and that vortex is a tornado—a spinning column of hot air rushing up and cooler air sinking down

GEEK MAMA FUN FACT!

Did you know there's a tornado in your tub? Next time you take a bath, pull the plug and watch. When the water gets to a certain level, you will see a tiny tornado coming from the drain. Why? It's a vortex. Air is coming up from the drain as water goes down, and they spin around each other.

❓ TAKE IT FURTHER

Can you make a tornado using only one bottle? Try it. Fill a bottle 3/4 of the way with water. Put a drop of dish soap in. Add a pinch of glitter and seal it up with the cap. Give it a swirl. What happens?

16

THE LEAK-PROOF BAGGIE OF WATER

What do you do with your sandwich bag and a sharp pencil after the first day of school? Save it for this science challenge: Poke a hole in a bag full of water and don't spill a drop. Think you can't? Well, you CAN! Practice this amazing experiment over the sink, or even outside, before you use it to dazzle your audiences.

STUFF YOU NEED

Water

Zipper-lock sandwich bag
  (It doesn't matter what size as long as it seals.)

Several sharp pencils

THINGS YOU DO

❶ Pour water into the bag. Fill it about ¾ full. Zip it. Make sure there are no leaks.

❷ Hold the bag from the top with one hand. With your free hand, grab your first sharp pencil and poke it through. Make the pencil stick out both sides of the bag. If not, there will be a mess. Any leaks? There shouldn't be one.

❸ Jab more pencils in. What happens? When you're done, hold the bag over the sink before you take any pencils out. Once you do, the bag will leak—a lot!

WHAT'S GOING ON?

It's not magic, it's science! It's all about polymers. Polymers are long chains of molecules that make up plastic. Plastic bags are made up of lots of long polymers. It's what makes those little bags so stretchy. When you jab a sharp pencil through the wall of the bag, the polymer strands move around so the pencil can slip through. These long chains then wrap back around the pencil making a seal that holds tight to keep the water in. Pull the pencil out and you have what scientists call "a big hole." The water will pour out. Try it!

 ## TAKE IT FURTHER

Try the same experiment with a water balloon. Does it work? What if you use something smaller, like a wooden skewer?

17

FLIP-BOOK ANIMATION

My boys have been fascinated by screens since they were able to focus on them. I spend a lifetime trying to come up with ways for them to be unplugged. Flip-books have been a favorite since they have been old enough to draw a line or a circle. Bringing a line or a simple shape to life with just pen and paper is a great way to empower a child and introduce the idea of what animation is. If they're watching cartoons, have them make them instead!

STUFF YOU NEED

Pad of sticky notes or a
   small notebook
Pencil

THINGS YOU DO

❶ Flip the pad of sticky notes with your finger from the bottom up. You are going to start drawing from the last page to the first. Start with a dot on a page near the bottom left corner. Flip a page down on top of it. Make the dot in the same space on this fresh page but make it slightly longer—the beginning of a line. It can go in any direction you choose.

❷ Flip the next page down and draw the same line in the same place but make this line even longer. Keep drawing over the last drawing and adding. Each picture needs to be slightly different than the last one to make the picture appear to move when it is flipped. Lots of pages showing very small differences will cause a slow movement; a few pages showing large differences will cause a fast movement.

❸ When finished, use your fingers to anchor the sticky notes on a table and use your thumb to flip the pages down. Watch!

WHAT'S GOING ON?

What you have done is create an animation. The drawings change slightly from one page to the next, so when you flip the pages, the pictures seem to move. Flip-books work because of an optical illusion caused by something called *persistence of vision*. That just means your eyes want to keep up with the flipping pages, but they can't. So instead of seeing individual drawings, your brain sees them as one connected drawing that looks as if it is moving.

 TAKE IT FURTHER

+ Keep experimenting with different drawings and have fun tricking your eyes and brain. Try making a circle that moves across the page, like a bouncing ball. Try drawing an egg that hatches into a bird. Maybe draw a circle that gets bigger and bigger and sprouts eyes, nose, and a mouth.

+ Want more optical illusion experiments? Try the one on page 65.

GEEK MAMA FUN FACT!

 Our brains create the illusion of movement not only when we're watching cartoons or movies—they actually do this all the time, such as when we blink.

(18)

HOW MANY PAPER CLIPS?

Here's a simple challenge—how many paper clips fit in a full glass of water? This science experiment explores the surface tension of water, and it might surprise you.

STUFF YOU NEED

Clear drinking glass
Water
Lots of paper clips

THINGS YOU DO

❶ Fill the glass of water as high as you can; don't let the water spill over the edges of the glass! Ask your child how many paper clips he or she thinks might fit into the glass of water without it spilling over.

❷ Drop a paper clip in. What happens? Drop paper clips in one at a time and count how many clips you can drop in.

❸ Take a look at the top of the glass of water. Can you see the water doming up? Keep going until the water finally overflows.

 WHAT'S GOING ON?

Water is sticky. Molecules of water like to hang out with each other. That's why the surface of the water bulges and forms a dome when you add the paper clips. The molecules hold on to each other so tightly that they keep the water from dribbling out. This is called *surface tension*. You can only push it so far, though. One too many clips can break the tension and the water will spill over.

If you have an eye dropper and a penny you can try this experiment. Use the dropper to place drops of water on a penny. Can you see the dome? How many drops can you add before the water spills off the penny?

19

DANCING PAPER SNAKE

Have you ever wanted to be a snake charmer? Frankly, real snakes, though I respect them, need to stay far away from me. But paper snakes are a different thing altogether. Even better than a cool paper snake is a paper snake that dances. What invisible force is making this paper snake dance?

STUFF YOU NEED

Paper plate or
    thin cardboard
Pencil
Crayons, colored pencils,
    or markers
Scissors
Tape
String
Lamp

THINGS YOU DO

❶ Cut the rippled edge off the plate.
❷ Draw the snake's head: Make 2 dots about 1 inch apart in the center of the plate with a pencil. Then draw a small triangle around the dots.
❸ Draw the snake's body: From the right corner of the triangle, draw a spiral radiating clockwise out to the edge of the plate. From the left corner, draw a spiral parallel to your first line out to the edge of the plate. Use markers, crayons, or colored pencils to decorate your snake. Use the scissors to cut the spiral snake out. [CONT.→]

④ Tape a piece of string to the top of the snake's head. Suspend the snake over a lamp. Make sure the lightbulb is one that gets warm (LED bulbs don't get as warm). You can also use a heated radiator. Make sure the paper is at least 4 or 5 inches above the surface. Wait. What happens?

 ## WHAT'S GOING ON?

The lamp is heating up the lightbulb. The lightbulb is heating up the air around it and making things move. Air is made up of molecules. When it gets heated up, the molecules move faster and bounce harder off each other. Warm air takes up more space. Cooler air pulls together; it takes up less space than warm air. Hot air is less dense than cold air, and therefore it rises. Cooler air sinks. The moving air spins the spiral snake.

 ## TAKE IT FURTHER

Want to try some more experiments with heat? Try The Ups and Downs of Hot and Cold on page 33. For cool experiments using density, check out page 11.

20

MOVE A PAPER CLIP WITH SOUND

Can you move a paper clip without touching it? You bet you can. You can move things with just the power of your voice!

STUFF YOU NEED

Plastic wrap
Bowl
Rubber band
2 paper clips
Your voice

THINGS YOU DO

❶ Wrap the plastic wrap over the opening of the bowl and secure it tightly with a rubber band. Make sure the surface of the plastic is tight—like a drum.
❷ Place the paper clips on the surface.
❸ Hum at the drum. Sing at the drum. Yell at the drum. What happens?

 WHAT'S GOING ON?

All sound is a vibration. When you hum, sing, speak, or yell, you are creating sound waves. These sound waves bump molecules of air together as they come out of you. Those vibrating air molecules bounce off the surface of your drum. When they bounce, they make the plastic bounce, and that moves the paper clips.

+ What happens if you sing high notes?
+ What happens if you sing very low notes?
+ If you sing loud, how does that affect the paper clip?
+ For more experiments with sound check out Straw Oboe on page 71 and Garden Hose Tuba on page 81.

GEEK MAMA FUN FACT!

Sounds funnel into your head through your ears. Sound waves bounce in and vibrate your eardrum and trigger movements that give your brain information. Your brain sorts it out, and you hear sounds.

(21)

WORLD'S HEAVIEST NEWSPAPER

We are walking under an ocean of air—and it's heavy! We have about 10 miles of air molecules over our heads at any given moment, and they are all pushing down on us all the time. The pushing down on us is called *atmospheric pressure*. You can see it in action in this experiment.

STUFF YOU NEED

⅛-inch-thick wooden ruler, yardstick, or meter stick (Note: It may break.)

Smooth-surfaced countertop or table in a clear area

Safety goggles

Newspaper

THINGS YOU DO

❶ Place the ruler on the countertop with about 3 or 4 inches jutting out over the edge. Give a quick "karate-chop" tap on the part of the ruler sticking out (i.e., not over the counter). The ruler should fly off (wearing safety goggles is a good idea). Place the ruler back on the counter with 3 or 4 inches jutting out.

❷ Take the biggest sheet of newspaper that you can find (the middle 2 pages that are attached work perfectly). Flatten out the sheet of newspaper as much as you can with your hand. Then place the sheet over the ruler on the counter.

❸ "Karate-chop" the ruler again, as hard as you can. Don't press the ruler down, but rather give it a sharp hit. The newspaper will hold the ruler down, even if you strike hard at the ruler. You may even split the ruler!

During the first chop, the ruler probably flew off the table and didn't break. During the second chop, the ruler probably didn't budge! It may have even broken. Why? Air pressure. When you spread out the newspaper on top of the stick, you basically created a large suction cup because you're preventing air from flowing underneath. When you strike the ruler, it tries to lift up against the newspaper, but because the air can't flow very quickly into the space between the table and the newspaper, most of it simply pushes down on the newspaper (and the ruler). If you push the ruler down (as opposed to tapping it quickly), some air will sneak in under the newspaper. Because there isn't a difference in pressure on the two sides, the newspaper will not keep the ruler down. That is why you need to hit the ruler quickly.

? TAKE IT FURTHER

For more experiments with atmospheric pressure check out Make Water Float on Air on page 35.

GEEK MAMA FUN FACT!

Our atmosphere is technically about 300 miles thick, but most of the air molecules are concentrated in a blanket about 10 miles thick. There is no hard edge to the atmosphere—it just gets thinner and thinner as you move out. When we think of the atmosphere we think of air, right? Air means oxygen. Well, oxygen is only a small part of air. It turns out that the atmosphere is made up of only 21 percent oxygen! The rest is composed of nitrogen, argon, carbon dioxide, varying amounts of water vapor, and trace amounts of many other gases. At sea level, the atmosphere exerts a pressure of 15 pounds per square inch (psi).

(22)

WALK THROUGH AN INDEX CARD

This is one of those impossible possibilities. With a little folding and cutting, you can walk through an index card.

STUFF YOU NEED

Index card
Scissors

Alternate cut lines from the folded edge and the open edge.

THINGS YOU DO

❶ Fold the card in half from short edge to short edge. Make a cut close to the top edge from the folded side to about ¼ inch from the open edge.

❷ Flip the card from front to back and make another cut parallel to the first, but this time from the open edge to about ¼ inch from the folded side. Continue making cuts like this, alternating sides.

❸ Now cut along the fold, skipping the pieces at each end.

❹ Unfold the paper carefully, and walk through!

WHAT'S GOING ON?

The alternate cuts actually convert the square area of the card into the perimeter of a larger rectangle. To visualize this, think of beginning a line of thin paper starting with the first fold at either end. The other cuts simply continue the line. You can trace the "line" of the paper with your finger.

TAKE IT FURTHER

+ Try different-size papers. What's the smallest paper you can make big enough to walk through?
+ Try a regular piece of printer paper. How big can you make the hole?
+ For more impossible paper science experiments, try Mysterious Mobius Strip on page 59.

23

MYSTERIOUS MOBIUS STRIP

A Mobius strip is another amazing impossible possibility. It is an object that has only one side and one edge. Sounds impossible, but it's not. And you can make one with paper, scissors, tape, and a simple twist.

STUFF YOU NEED

One 8½ x 11-inch sheet of printer paper
Scissors
Tape

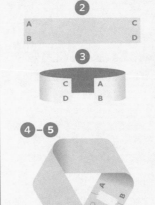

THINGS YOU DO

❶ Cut a strip of paper about 2 inches wide from the long edge of your paper. It should measure about 2 x 11 inches.

❷ Place the strip on the table in front of you so the long side of the strip is positioned horizontally. Write the letter "A" at the top-left corner of this strip, the letter "B" at the bottom-left corner, the letter "C" at the top-right corner, and the letter "D" at the bottom-right corner.

❸ Now make a ring by placing the back of the letter D to the front of letter B. The back of the letter C will be on top of the front of letter A.

❹ Holding the AB side in your left hand, take the CD side and twist it so that the front of the C side is [CONT.→]

against the front of the B side and the front of the D is against the front of the A.

❺ Tape the twisted strip of paper in place. You should have 1 long twisted loop that has only 1 side and 1 edge!

 WHAT'S GOING ON?

You'll find that if you keep your finger moving along an edge, you'll end up touching every edge of the object and end up right back where you started—which means that this object has only one edge. Imagine you are an ant crawling along. You will crawl over the whole strip on both sides and come back to the place you started. It's so interesting to actually see something that seems impossible.

 TAKE IT FURTHER

Try to cut your strip in half. Cut right down the middle, like you are cutting two thinner strips. Bet you don't get two! You end up with a much larger, thinner Mobius strip.

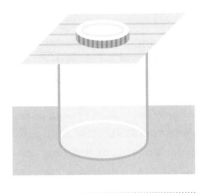

24

COIN DROP

This science experiment is so fun and seemingly impossible that it feels more like a magic trick. You won't end up doing it only once. It's fun and addictive. And it packs a very big science punch—even if the kids are more consumed with just doing it over and over and over and over. They will be experiencing the joys of gravity, inertia, and friction. You don't have to get bogged down in the words, but the concepts are something they will see again and again. It still surprises and delights me every time!

STUFF YOU NEED

Index card
Drinking glass
Quarter

THINGS YOU DO

❶ Place the card on top of the glass. Make sure there is enough space to give an edge of the card a good flick without smacking your finger on the glass.

❷ Place the quarter on top of the card directly in the center so it rests over the cup's opening.

❸ Make a hypothesis about what will happen when you flick the card. Now do it! Flick the edge forward in a plane parallel to the tabletop. (Don't flick the card up from underneath.) Where did the coin go? *KaCHING!* It should have dropped straight into the glass!

WHAT'S GOING ON?

Thank Sir Isaac Newton for this little science gem. He came up with a bunch of laws about how the world works. This experiment shows Newton's First Law of Motion, which can be summed up like this: *An object at rest will stay at rest unless an outside force acts upon it. An object that is moving will stay moving until something stops it.* In the case of our experiment, the coin is happily sitting at rest on the card on the glass. It doesn't do anything until you flick the card out from under it. When you do that, gravity (an outside force) acts upon that resting coin and yanks it down into the glass. The coin drops—it is in motion. And it stays in motion until it is acted on by an outside force—in this case—the bottom of the glass, which stops it. *Boom!* Newton's First Law of Motion.

Are you wondering why the coin doesn't take off and fly across the room with the card? That's a question of friction. Friction happens when two things rub together. If you flick the card right, it slides out so quickly from between the glass and coin that it doesn't create enough friction to pull the coin with it. The force of gravity is way stronger than the force of friction in this case.

 TAKE IT FURTHER

+ Does this work with other coins?
+ Does it work with a stack of coins?
+ What other objects will work? Raisins? M&Ms? Marshmallows? Dried pasta? Make a chart if you want to.

(25)

PAPER FOLDING CHALLENGE

Seems simple enough, right? Fold paper. It's easy. Well of course it is. But science says you probably can't do it more than seven times. Don't believe it? Try it!

STUFF YOU NEED

One 8½ x 11-inch sheet of paper

THINGS YOU DO

❶ Lay the paper flat on the table. Fold it in half. Make a sharp crease with your nail. Now fold that piece in half again (that's two times). Do it again (three times). And again (four times).

❷ How many times can you fold the paper?

 WHAT'S GOING ON?

The number of layers of paper doubles with each fold. So you start with a single layer, then you have two layers, then four, then eight, then sixteen, then thirty-two, then sixty-four layers after six folds. Maybe if you are very strong, and you use a pair of pliers, you can get to seven folds and 128 layers, but it probably won't be pretty. After that, the sheet of paper is so tiny and thick that it has too many layers and there is no way to fold it again. The fibers that make up the paper can no longer bend.

If you're having fun folding paper, don't stop! Try A Great Paper Airplane on page 120 and Paper Wheel Wind Whirler windmill on page 123.

GEEK MAMA FUN FACT!

What if you have a huge sheet of paper to start with? In episode 72 of the Discovery Channel's 2007 *Mythbusters* series, the Mythbusters guys took a football-field-size piece of paper and used a forklift and a dry-roller to make eleven folds! That's 2,048 layers, before they reached the limitations of folding.

26

DISAPPEARING RAINBOW

Usually you find yourself looking for a rainbow. But in this experiment you start with a rainbow and make the colors disappear right before your eyes! All you need is a little spin. Now you see it. Now you don't!

STUFF YOU NEED

Pencil

Jar or cup with at least a 4-inch-diameter opening

White paper plate (This experiment works best with a thicker plate.)

Scissors

Crayons or markers in all the primary colors (blue, yellow, and red)

Crayons or markers in all the [CONT.→]

THINGS YOU DO

❶ Use the pencil to trace the mouth of the cup onto the plate to make a smaller circle. Cut out the smaller circle.

❷ Draw a line through the middle of the circle to make 2 halves. Next, divide each half into thirds by drawing an X through the whole circle. You should have 6 equal pie slices.

❸ Color each pie-slice section of the plate a different color of the rainbow: red, orange, yellow, blue, green, purple. [CONT.→]

**secondary colors (pur-
ple, orange, and green)**
3 feet of string

❹ Poke 2 small holes in the center of the circle, about ½ inch apart. Thread the string through the holes and tie the ends in a knot.

❺ Hold one end of the string in each hand with the disc in the middle. Now spin it! One-person method: Wind up your disk by swinging the disc like a jump rope, then move your hands closer and farther apart as it spins faster and faster. Two-person method: One person holds the string with the disk in the middle, and the other spins the disc to wind it up. When the string is wound up tight, pull both hands away from each other so that your colored disk spins quickly.

❻ Now look. As the wheel spins, what do you notice about the colors? They should begin to blur together and, eventually, appear to disappear and turn white!

 WHAT'S GOING ON?

In the whirl, all your eyes can tell you is that they see white! Why do the colors disappear? Technically they actually don't disappear. They simply blend together. When all the colors of the rainbow mix the result is white light. All the colors are still there, but because of the spinning disc, our eyes can't see the separate colors anymore. They blend into white. Think about it: rainbows start out as white light from the sun, then they split into different colors when they bump into raindrops. You are just doing the opposite with this experiment.

+ Try making another plate with two equal sections: blue on one half and yellow on the other. What happens when you spin it? Why? Try red and yellow. Try blue and red.

+ Looking for more optical illusion experiments? Try Flipbook Animation on page 46. Want more color? Check out Fizzy Color Mixing on page 102.

GEEK MAMA FUN FACT!

Rainbows happen when sunlight shines through water droplets. The water droplets break up the sunlight into the different colors that make up the white light. It's always in the same order: red, orange, yellow, green, blue, indigo, violet. It has to do with wavelength. Violet light has a shorter wavelength, and red light has a longer wavelength. I tell my kids about ROY G. BIV— you can use his name to remember the order of the colors.

ON THE PLAYGROUND

LOUD, MESSY, AND FREE-RANGE EXPERIMENTS

Go outside! Remember that? That's what my mom would say to us when my brother and I were kids. "Get out and have adventures." We'd roll our eyes and dutifully oblige her and then get lost in the amazement of some awesome experiment or adventure. I remember the first time we dug a hole—there were layers of different kinds of dirt. Cool! I remember the first time we hit rocks with a hammer and split them. There were sparkly bits inside! Wow! When my boys were small, we got started early with the notion that outside is wonderful because you can be loud outside. You can be messy! And you can explore and adventure.

Here are a few of our favorite experiments and explorations best suited to the great outdoors. You might get wet. You might get dirty. You might get loud. You might make a mess. At least *hopefully* you will!

27

STRAW OBOE

This straw oboe is more than just a noisemaker—though that part is superfun. It's also a great way to start to understand sound and music. With a simple straw and a pair of scissors—and a strong set of lungs!—you can learn about vibrations and get a good sense of how reed instruments work.

STUFF YOU NEED

Plastic drinking straw
Scissors

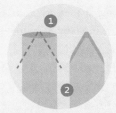

When you cut the corners of the straw tip, it should look like the shape of a blunt pencil.

THINGS YOU DO

❶ Pinch one end of the straw flat. Put the straw on the table and crease the flattened end at the tip with your nail. It won't stay flat, but when you look at it from above, it will look like a football shape.

❷ Cut the two corners off the flattened end. You should end up with a shape that looks like a blunt pencil.

❸ You're done! Now play it! Put the cut end in your mouth just past your lips and blow. You may need to move it in and out until you find [CONT.→]

the place where the two flaps of the straw vibrate. Be patient! Keep trying. When you get it, blow hard!

 WHAT'S GOING ON?

You have made a double reed instrument. By creating the two flaps, you can make the straw vibrate against itself. This is how reed instruments work. When you blow into them, the reed vibrates, the vibration makes a sound, the air inside the instrument vibrates, and the sounds are amplified.

 TAKE IT FURTHER

+ Want to make different notes? Try cutting small diamond-shape holes down the straw. Cover them up with your fingers and blow for low notes. Take your fingers off the holes, and you shorten the column of vibrating air. Shorter vibrations have higher sounds.

+ Want to make more noise? Try the Bee Hummer Buzzing Paper on page 74 or the Screaming Balloon on page 77. Want to make more instruments? Check out Garden Hose Tuba on page 81.

GEEK MAMA FUN FACT!

Woodwind instruments make sounds when you blow air into them. They used to all be made out of wood. That's why they're called wood-winds. Wood plus wind. Nowadays woodwinds are made out of metal, plastic, and wood. Woodwinds are separated into two categories—flute instruments and reed instruments. Flute instruments include flutes, piccolos, and recorders. They make sound when you blow air across an edge in the instrument. Reed instruments have a thin wooden reed, or two, that vibrates when air passes over it. Clarinets, oboes, bassoons, saxophones, and bagpipes are reed instruments.

28

BEE HUMMER BUZZING PAPER

This project uses the power of vibration to make a cool instrument that sounds like a swarm of buzzing bees when you spin it around. It's a win-win experiment because it's fun to make, fun to spin, and the sound is creepy-fun!

STUFF YOU NEED

Pencil erasers (the kind that fit over the top of a pencil)

Wooden craft stick

Index card

Scissors

Stapler (Make sure it's strong enough to staple the index card to the craft stick.)

2-foot-long piece of string [CONT. →]

THINGS YOU DO

❶ Stick an eraser on each end of the wooden craft stick, then place the short edge of the index card along the length of the stick. (You may have to trim the card a bit to fit it between the erasers.) Staple the card in place; I like to use at least 3 staples.

❷ On one end of the stick, tie the string in a knot right next to an eraser. Then stretch the fat rubber band around the whole thing, eraser to eraser. Make sure the string is coming [CONT.→]

¼-inch-wide rubber band (It should be long enough to stretch lengthwise around the wooden craft stick.)

out on the top between the two sides of the rubber band.

❸ Cut about ⅓ off the bottom of the index card.

❹ Make the paper buzz by swinging the whole thing around in a circle by the string. Can you hear it buzz?

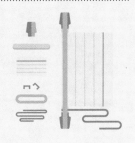

 WHAT'S GOING ON?

When you spin the hummer, moving air causes the whole thing to vibrate—the index card and the rubber band. The vibrations make a sound, just like the vibrating strings on a guitar or a violin or a ukulele make a sound. The paper vibrates, too, and it makes the sound of the rubber band vibrations louder.

 TAKE IT FURTHER

+ What happens to the sound when you spin it slowly? What happens if you spin it faster? (Slower spins make slower vibrations and lower sounds. Faster spins make higher vibrations and higher sounds.)

+ What happens if you change the shape and length of the index card?

+ Want to make more vibrations? Try the Straw Oboe on page 71 or the Screaming Balloon on page 77. Want to see how vibrations can move something? Try Move a Paper Clip with Sound on page 52.

(29)

SCREAMING BALLOON

Halloween is big in our house, and all things Halloween are appreciated. But who needs Halloween just to make a balloon scream? Any day is a good day to try that (outside)!

STUFF YOU NEED

Black marker

Balloon (I like to use a big white one.)

Hex nut (Hex nuts have 6 sides—count 'em.)

THINGS YOU DO

❶ Use a marker to draw a face on the balloon. Don't inflate it; draw it while the balloon is small and flat. Wait for the ink to dry. When it's dry, put the hex nut in the balloon and shake it down.

❷ Blow up the balloon. Watch the face grow! Don't blow it up too much. That will make the walls more fragile, and it will pop easily. Tie off the balloon.

❸ Have your child grab the balloon at the knotted end and swirl it in a circle. The hex nut will bounce a bit, and then it should roll around inside the balloon. Be patient! Soon you will hear the ghost balloon screaming! Next, you will be screaming with joy—the sound is oh-so pleasant!

 WHAT'S GOING ON?

A couple of things are happening in this experiment. First, the hex nut is whipping around inside the balloon and making the walls of the balloon vibrate. Vibrating things make sounds. This one is a sound like screaming. Another thing that is going on is about motion. The hex nut would rather go in a straight line, but the inside walls of the balloon are stopping it from doing that, so it has to whip around in a circle. The balloon is imposing a force called a *centripetal* force. Centripetal force is the inward force on a body that causes it to move in a circular path. It's the same force you feel on an amusement ride that spins. You want to go in a straight line, but the seat of the ride stops you and you spin around. And that can cause a whole different kind of screaming!

 TAKE IT FURTHER

+ What happens if you spin it faster? Slower?
+ What if you have a bigger balloon? A thicker balloon? A smaller hex nut? Why? (Hint: It's all about vibrations. Faster vibrations sound higher, and slower vibrations sound lower.)
+ Want to make more noise? Try the Bee Hummer Buzzing Paper on page 74.

(30)

JUICE GLASS XYLOPHONE

You and the kids are dying to play the xylophone! Who isn't?
But you don't have one! No xylophone, no vibephone, no
glockenspiel? No problem. What can you do? You can make
your own! You can do this right at the kitchen counter. Add food
coloring, and the whole thing becomes a rainbow of sound.

STUFF YOU NEED

8 juice glasses, clean
and dry (You can use
jam jars, too, as long
as they are all the
same size.)
Chopstick or pencil
Water
Food coloring

THINGS YOU DO

❶ Arrange the glasses in a row. Have
your kids tap each glass with a chopstick
or pencil to be sure the glasses all start
out with more or less the same note.

❷ Leave the first glass empty and pour
some water in the second glass. Have
your child tap that glass. How does the
sound differ? Here's the fun part: pour
water in the subsequent glasses and

tap until you get what sounds like a scale. It won't be perfect, but you
will hear the changes.

❸ Turn your xylophone into a rainbow. Decide which colors will go
where. (Personally I like to place the colors in order of an actual
rainbow—ROYGBIV—red, orange, yellow, green, blue, indigo, violet.)

❹ Play! Gently strike the glasses with a chopstick or pencil.

WHAT'S GOING ON?

When you strike the glass you create vibrations—or sound waves. Sound waves travel through the liquid in the glasses. Change the amount of water, and you change the sound of the pitch. The glasses with less water have shorter, faster vibrations and higher pitches. The glasses with more water have slower vibrations and lower pitches.

 TAKE IT FURTHER

+ Write music basing the notes on color. Tap out "Hot Cross Buns," for example, and see which colors match each note. Then use crayons to make a line of circles that represent notes. Make your own compositions, then see if you can repeat them by looking at the color code.
+ Try tapping a glass in a different spot. How does the sound change when you tap closer to the top?
+ What happens when you tap the glasses with a metal spoon?
+ Different liquids have different sounds. Try using things like vegetable oil, milk, ketchup, and syrup to see how each one affects the sound waves.
+ Try making xylophones with different materials. One of my favorites is a Wrench Xylophone: Get a set of five wrenches of different sizes. Place them in order by size on the back of an egg carton—they should fit in between the egg-shaped bumps. Now play!
+ Want to make more instruments? Check out Straw Oboe on page 71 and the Garden Hose Tuba on page 81.

31

GARDEN HOSE TUBA

Haven't you ever found your-self wondering, "Gosh, I have this old hose that I would love to recycle, but how?" Here's how! Make it into a tuba. Instead of sprinkling water into the garden, you will sprinkle lovely music into the neighborhood!

STUFF YOU NEED

3-foot section of garden hose (If it's too long, it will be hard to blow air through it.)

Duct tape

Funnel or plastic bottle

Plastic snap-on hose connector

THINGS YOU DO

❶ Make the shape of your horn by coiling the hose into a loop where the two ends overlap and point in different directions. Tape the hose together at the overlap.

❷ Insert the small end of the funnel into the hose. Tape it securely in place. If you don't have a funnel, use a 2-liter pop bottle. Cut the bottle just under the shoulder where the neck slopes into the body of the bottle. It should look like a funnel. Place the small end into the hose, and tape it securely.

❸ Add the mouthpiece. You can simply try playing the [CONT.→]

instrument without a mouthpiece, but if you have a snap-on plastic hose connector, it can make a very comfy mouthpiece. Insert the smaller end into the hose at the other end from the funnel. Tape it securely.

❹ Play your horn! Vibrate your lips together to make a buzzing sound. Place your lips at the mouthpiece and buzz them again. Use a lot of air. To change notes, tighten or loosen your lips as you buzz them.

❗ WHAT'S GOING ON?

This is how brass instruments work. You buzz your lips into a mouthpiece. The vibrations created between your lips and the mouthpiece travel down the long tube—rattling a column of air molecules. Faster, shorter vibrations create a higher pitch. Lower notes come from looser lips and slower, longer vibrations. It used to be all horns were played this way—using the tightening and loosening of lips to change pitch. But then things like valves and slides were introduced to change the length of the column of air. In a trumpet, if you open a hole in the tube, you get a higher note because the column of air is shorter. In a trombone, if you lengthen the slide the column of air is longer and you can get lower notes.

❓ TAKE IT FURTHER

Try different kinds of tubes you can find around the house, such as wrapping paper tubes or paper towel tubes. How does the sound differ?

32

CORNSTARCH QUICKSAND

This experiment has to be done to be believed. I can't tell you how many times we have done this on a summer day outside—it can be messy. (I did do this experiment once in a school library as an assembly, and I think I made the librarian cry by accident. Outside is good.) You take a simple kitchen ingredient, add water, and get a goo that sometimes acts like a liquid and sometimes acts like a solid. It still thrills and amazes us all.

STUFF YOU NEED

One 16-ounce box cornstarch
Kitty litter box or dishpans
1 to 2 cups water

THINGS YOU DO

❶ Head outside or to a place where you can make a mess safely. Pour the entire box of cornstarch into the box or pan. You'll want to stick your hands into this snow-white powder and see how it feels. Squish it. Rub it. What does it feel like? What does it sound like? It's such an interesting texture.

❷ When you are ready, add about 1 cup of water to the cornstarch. Mix the goo. At first it may feel rock hard and too dry and like cement. Be patient. If you need to add a little more water, do it slowly. The mixture should be the consistency of slightly runny yogurt or honey.

❸ Explore the goo. Gently place your hand in the mixture. Moving slowly, you can stir the mixture and it feels like runny yogurt. [CONT.→]

Now try mixing it fast. What happens? Grab a handful of the goo. Hold it in your hand and watch as it runs right through your fingers. Grab another handful and try rolling it between your palms to make a ball. Stop rolling it and watch it turn to liquid and drizzle away. Try slapping the top with the flat part of your hand. What happens?

❹ When you are finished, pour the glop into a zipper-lock plastic bag. Seal it up, and keep it for playing later. (The contents may separate out a bit; just mix it up again and it should work fine.)

 ## WHAT'S GOING ON?

When you stir slowly, the mixture acts like a liquid. But when you manhandle it, it acts like a solid. Why? Technically, this is called a *non-Newtonian fluid* because it just doesn't act like other fluids. The cornstarch (which is a solid) is suspended in the water (which is a liquid). The cornstarch molecules slide past each other easily when the stirring is slow. But stir too fast or put too much pressure on the mixture, and the molecules line up and won't slide past each other anymore—they act like a solid.

In nature, or at least in old-time jungle movies, quicksand is an example of this kind of suspension mixture. Quicksand is a soupy mixture of sand and water, where the sand is literally floating on water. Move slowly in the mixture and the sand can slide past itself. But thrash around and move too fast and the sand particles will lock up and suck you down. If you ever do get caught in quicksand, relax. You will float, and then you can gently swim to shore.

TAKE IT FURTHER

Toss a small plastic figure into your quicksand. What happens?

(33)

EXPLODING SIDEWALK CHALK

Whenever I show up in classrooms the kids gleefully ask, "Are you going to blow stuff up?" and the grown-ups cautiously ask, "Are you going to make a mess?" *Yes* is always my go-to answer. This is a favorite experiment because you do get to blow stuff up and you do get to make a mess and it's all great science.

STUFF YOU NEED

Zipper-lock sandwich
    bags
Corn starch
White vinegar
Food coloring (We like
    using neon food colors
    because they're so
    bright and fun.)
Baking soda
Tissues

THINGS YOU DO

❶ Set up your chalk bags. Put ¼ cup of cornstarch in a zipper-lock bag. Make sure to get the cornstarch *in* the bag and not on the edges of the bag. If the zip gets gunky, it won't hold and the explosion will be much more of a fizzle.

❷ Add ¼ cup of vinegar to the cornstarch in the bag. Zip the bag and mix the ingredients using your fingers on the outside of the bag. What you want is a thick liquid consistency—like honey or ketchup.

❸ Open the bag and add a few drops of food coloring. Zip it shut and mix some more. Make a bunch of bags with different colors!

❹ Set the bags aside. You will need to remix them just before you set them up because the cornstarch can separate out and harden a bit. A quick remix does the trick. [CONT.→]

❺ Make baking soda burrito bombs. Open up a tissue and make sure it is only one layer thick. Tear the tissue in half. Put a heaping tablespoon of baking soda in the middle of the tissue half. Fold the long edges in and roll from a short edge to make a burrito with the baking soda inside. (This gives the soda a slight barrier to the liquid so you will have some time to plop the bomb in and zip the bag before any reaction takes place.) Make enough burrito bombs so you will have one for each bag.

❻ Explode your chalk! Find a good space to explode your chalk bags, such as the street, sidewalk, or driveway. They will leave a chalky mark, but it will wash easily away with a hose or a good rain. Line up your bags. Open one. Plop a baking soda burrito inside. Zip the bag shut FAST! Give the bag a little shake. Place the bag down with the zipper side facing in the direction you want your chalk to flow. Stand back and watch! The bag should inflate and inflate and *POP!* A lovely, chalky goo will flow.

 ## WHAT'S GOING ON?

This is just another riff on the old "add an acid (vinegar) to a base (baking soda) and make a chemical reaction" experiment that takes the homemade volcano even further. The cornstarch and the vinegar with the food coloring make the chalky acid mixture. The baking soda is the base, and when the two of them collide— watch out. A chemical reaction is produced. The product of that reaction is carbon dioxide gas, which expands quickly. It pushes against the inside of the bag until it reaches a breaking point, and *BOOM!* The gas escapes, and the chalky mixture does, too.

The beauty of this experiment is that you can see the gas produced in the reaction as it inflates the bag. The satisfying explosion belches out a colored goo that dries and brightens up the street or sidewalk until a good rain comes along.

❓ TAKE IT FURTHER

+ Try designing with the explosions. Be intentional in placing the bags on the street.
+ After the bag explodes, use your fingers and paint with the chalk.
+ When the chalk explosions dry, take regular sidewalk chalk and turn your blobs into new worlds. We have made fantastic undersea worlds and extraterrestrial planets as well as monsters and microscopic wonders.
+ If your child likes the fizzy part of this experiment, try Fizzy Color Mixing on page 102. If it's the explosive part, try Pop Stick Explosion on page 99.

MAKE YOUR OWN SIDEWALK CHALK

If coloring the sidewalks is something your kids find inspiring, you can whip up a batch of your own sidewalk chalk.

Gather some wrapping paper tubes to use as your molds (these are a good diameter—slimmer than toilet paper tubes—but you can use any tube you like). Cut the tubes into 6-inch segments and line the inside with wax paper. Tape the bottoms closed.

For each stick of chalk, mix 2 to 3 tablespoons of tempera [CONT.→]

paint with ½ cup of water. Stir in ¾ cup of plaster of Paris. Spoon the mixture into a small plastic bag. Snip off the bottom corner of the bag and squeeze the goo into the mold. Repeat the process for however many colors you'd like to make. Make sure the tube stays propped up. Let it dry for at least 12 hours. Peel off the mold and draw!

MOON SAND

Of course sand is a favorite thing to play with. You can push it around, dig in it, and even build stuff. As far as scientific explorations, it's perfect for looking at how things flow. You can talk about solids and liquids.

You can see how sand takes the shape of its container. Each particle is like a molecule. It's a great way to get kids thinking. But MOON SAND? Well that's even cooler still because it can also hold its shape. Plus, you get to make a concoction that is superfun to play with. And it smells good, too.

STUFF YOU NEED

4 cups sand (You can get it at the beach or buy it at a hardware store.)

Plastic square dishpan or unused litter box or bucket

2 cups cornstarch

Yellow food coloring, liquid [CONT. →]

THINGS YOU DO

❶ Place the sand in the container. Add the cornstarch and mix well. Talk about how it's a mixture because you can see the parts.

❷ Put a few drops of food coloring in the warm water. Add a few drops of lemon extract and the drop of dish soap (the soap will help prevent the growth of bacteria over time). [CONT.→]

watercolors, or pow-dered paint (optional) 1 to 3 cups warm water Lemon extract (optional) Drop of dish soap	❸ Add the colored lemony water to the sand mixture and stir the whole thing up. Start with 1 cup and add more until you get the consistency that you want for your moon sand. ❹ Play! It's a sand and a clay.

 WHAT'S GOING ON?

The cornstarch holds the moisture and keeps the sand damp, allowing it to stick to itself better. That means you can sculpt and smush the sand and it can hold its shape. This experiment is a good springboard for talking about states of matter, if you are so inclined. A solid holds its shape (think ice). A liquid takes the shape of its container (think water). You can see these traits with the sand. It acts like a liquid because it takes the shape of its container. The particles flow past each other. It can also act like a solid because the particles can stick together and it can hold its shape.

 TAKE IT FURTHER

+ Don't stop at lemon! In the fall, use pumpkin pie spice and orange food coloring! In the winter, use vanilla flavoring and no color or pine scent and green color. In the spring, you can add any essential oil of flowers.
+ Freeze your sculptures and see what happens.

(35)

OBEDIENT BUBBLES

This activity has double the fun. You make bubbles and you use balloons to make the bubbles move. It's fun and it's science, too. While you play, you are seeing firsthand how the power of static electricity and magnetism make bubbles do your bidding!

STUFF YOU NEED

Balloon
Bubble Goo (see recipe below) or soap bubbles

THINGS YOU DO

❶ Blow up your balloon and tie it off. Charge your balloon with static electricity by rubbing it on your head until you hair stands on end, or rubbing it on your sweater.

❷ Blow the soap bubbles. Bring the charged balloon close to the bubbles and watch them move toward the balloon. Try moving the bubble around the room with the balloon.

 WHAT'S GOING ON?

When you rub a balloon against your hair or sweater, you build up electrons, which creates a magnetic charge on the balloon. Electrons are negatively charged particles. These negative particles are attracted to the positive particles on the surface of the bubbles. When you get them close to each other, the bubbles are drawn to the balloon.

Make a bubble obstacle course and see how far you can get with your "trained" bubbles!

MAKE YOUR OWN BUBBLE GOO

A bubble is just air wrapped in soap film. Soap film is made from soap and water. The outside and inside surfaces of a bubble consist of soap molecules, and a thin layer of water lies between the two layers of soap molecules, sort of like a water sandwich with soap molecules for bread. They work together to hold air inside. Make some easy bubble solution and wands right at home for bubbly fun.

Pour ½ cup of dishwashing liquid into a container (you'll want to use a container with a lid; large yogurt containers work well). Add 4 tablespoons of glycerin (you can find glycerin at drugstores or craft stores) and 4½ cups of water. Put on the lid and give the whole thing a shake. Open it up and let it settle.

Bend a pipe cleaner into a loop and twist it shut. Dip the loop in the bubble goo and blow.

(36)

SUPER-SQUIRTER WATER BLASTER

All you need to make a super squirter is a few household items and some atmospheric pressure! It's easy. In fact, it's a blast! Words of advice: Do this one outside on a hot day. It's even more fun if you make more than one and share with a friend.

STUFF YOU NEED

Hammer
Nail to make holes
Plastic pop bottle (the bigger the better)
Balloon
Water

THINGS YOU DO

❶ Hammer a nail through the cap of the bottle. Pull it out and set the cap aside.

❷ Using the nail again, make a hole near the bottom of the bottle on the side. Take the balloon, dangle it inside the bottle, and stretch the mouth of the balloon over the mouth of the bottle.

❸ Fill the balloon with water. While the water is still flowing, and when the balloon is full, place a finger over the hole in the bottle. Turn off the water. Carefully place the cap on the bottle and tighten it. Don't take your finger off the hole in the bottle!

❹ Make sure you are in a space that's okay to get wet. Ready? Aim! Take your finger off the hole!

WHAT'S GOING ON?

The hole in the bottom of the bottle allows the balloon to be filled with water. The balloon squeezes all the air out of the bottle. When you plug the hole with your finger everything is in equilibrium. The pressure on the outside is the same as on the inside. When the hole is plugged, the bottle protects the balloon from being acted on by air pressure. When you take your finger off the hole in the bottle, air pressure pushes into the hole in the bottle. That air pushes on the rubbery walls of the balloon and squeezes the water. The water has to go somewhere, so it shoots out the hole in the cap. Boom! Squirt power—with a little help from the force of air pressure.

 TAKE IT FURTHER

+ Hold a contest—who can squirt the farthest?
+ What if the hole in the cap is bigger?
+ What if the hole in the bottle is bigger?
+ For more atmospheric pressure exploration, try Make Water Float on Air on page 35 and World's Heaviest Newspaper on page 54.

STRAW
SPRINKLER

You don't have a sprinkler?
Make one! Using a straw,
a shish-kebob stick, and a
little science, you can move
water up and out! On a hot
day, this is such a fun way to
cool off and play with a little

atmospheric pressure while you are at it.

STUFF YOU NEED

Plastic straw
Marker
Ruler (optional)
Wooden skewer
Scissors
Tape
Bowl or bucket of water
   (A kiddie pool is great,
   too!)

THINGS YOU DO

❶ Find the center point of the straw
and make a dot with the marker.
From that point, measure (with a
ruler, if you want) about 1 inch away
on each side and mark those spots.
❷ Poke your skewer through the
center dot. Move the skewer through
the straw so that it extends about
4 inches from the straw. Hold the
skewer with the point up. Using scis-
sors, carefully snip only ¾ of the way through the straw at the other
marks from the bottom up.
❸ Fold the straw up where the cuts are. The tip of your [CONT.→]

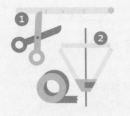

triangle will be pointing toward the pointy end of your skewer. Use the tape to hold the ends down—but don't cover the openings of the straw.

❹ Place the pointy side of the triangle underwater. Make sure the other two corners of the triangle are out of the water. Twist the skewer. (You can put it between your palms and rub back and forth to twist for a real spray!) Are you wet yet?

❗ WHAT'S GOING ON?

When you spin your contraption, you are spinning the water inside it as well. The water wants to push outward (think salad spinner). The technical term for the pushing outward when spinning is called *centrifugal force*. The thing is, the water in the straw can't push outward for long because it is in a straw. It bumps up against the walls of the straw. That doesn't mean it stops, though. The sides of the straw make the water actually go up—it's the only direction available. If you spin the straw fast enough, then water will get pumped up the straw and whip out of the holes in the straw at the corners of the triangle.

❓ TAKE IT FURTHER

+ Try using different length straws to see how far you can spray.
+ Try using different containers of water. Does a deep bowl work better than a cup?

(38)

POP STICK
EXPLOSION

Explosions play big in our
house. Here's one you can
do anywhere. All you need
is a handful of wide wooden
craft sticks or tongue depressors. When you weave them
together under pressure, you
spring-load the whole structure. Undo it and you get an exciting
chain reaction blast.

STUFF YOU NEED

**At least 8 wide wooden
craft sticks or tongue
depressors**

THINGS YOU DO

❶ Start weaving the craft sticks
together. Place stick 1 straight up
and down on a flat surface. Place the
bottom end of stick 2 on top of the
bottom end of stick 1 at a 45° angle.
Cross stick 3 over stick 2 to make an
X; place the middle of stick 3 under
the middle of stick 1.

❷ Stick 4 is your first weave. Place
stick 4 parallel to stick 3. Place the
center of stick 4 over the top [CONT.→]

half of stick 1; place the bottom end of stick 4 under the top half of stick 2.

❸ Place stick 5 parallel to stick 1. Place the bottom end of stick 5 under the top end of stick 3; place the middle of stick 5 over the top half of stick 4.

❹ Place stick 6 parallel to sticks 4 and 3. Place the bottom end of stick 6 under the top end of stick 1; place the middle of stick 6 over the middle of stick 5.

❺ Place stick 7 parallel to sticks 1 and 5. Place the bottom end of stick 7 under the top of stick 4 and over the middle of stick 6. Keep going, following this weaving pattern, for as long as you want.

❻ When you are done, pin your weave in place with a holding stick. Tie a string around the center of this stick if you like. That way you can explode your sticks from a distance. The holding stick will lie over the last stick you placed. Both ends will be tucked under in the pattern. Wedge it in tightly to keep it in place.

❼ Explode your sticks! When you are ready, remove the holding stick. Stand back! The chain reaction makes the sticks fly up together before exploding.

WHAT'S GOING ON?

When you weave the sticks together, you are creating the chain and putting pressure on the sticks. The energy is stored in the sticks. They are just dying to fly away from each other, but they are stuck. When you unstick one, the energy is released in the first weave, and then the next one goes, and then the next in a fast chain reaction.

 TAKE IT FURTHER

+ How long a chain can you make?
+ Try coloring the sticks different colors and making patterns before you explode them.

(39)

FIZZY COLOR MIXING

This one is a twofer—you get two bangs for your buck. Color mixing is fun on its own, but everything is better with a fizz and a chemical reaction, right? My favorite science experiments are the ones that overlap with art, life, language, and wonder. This is a simple one, but it's so cool the kids will want to keep playing. And they may just walk away with a few light and color concepts in their tool box.

STUFF YOU NEED

6 plastic cups, plus some extras for experimenting (Little yogurt cups are perfect!)
1 cup vinegar
Food coloring
3 eye droppers, pipettes, or straws
1 cup baking soda

THINGS YOU DO

❶ Set up your acid station. Set out 3 cups with about ⅓ cup of vinegar in each. Make them primary colors by adding food coloring: 2 drops of blue to the first, 2 drops of red to another, and 2 drops of yellow to the last. Place an eye dropper in each cup.

❷ Set up your base station. Set up 3 more cups with ⅓ cup of baking soda in each. Make them into primary colors by adding food coloring: 3 drops of blue to the first, 3 drops of red to another, and 3 drops of yellow to the last.

❸ Set out a few extra empty cups for experimenting.

❹ Take a dropper full of 1 color of vinegar and drip it into another color baking soda. What happens? What colors do you get? [CONT.→]

How do you get green? Purple? Orange? See what kinds of colors your kids get while they fizz the day away.

 WHAT'S GOING ON?

The fizz happens because when you mix an acid (vinegar) with a base (baking soda) you get a chemical reaction that produces gas—or in this case, fizz. But the real fun is also in the color mixing. You can make any color in the rainbow if you have primary colors. Primary colors are red, blue, and yellow. You can't get primary colors by mixing, but you can use them to make other colors. This is such a fun, open-ended experiment that kids will create different potions of colors and see how they combine.

 TAKE IT FURTHER

+ Try mixing your base colors with enough water to make a paste and then freeze them.
+ If the kids like fizz and color, try the Exploding Sidewalk Chalk on page 87.

ICE CUBE TOWER

Try and stack ice cubes. Go on, try! What happens? They slide right off each other. But there is one common kitchen ingredient that you can sprinkle on to make the ice cubes freeze in their tracks and hold fast. Go ahead and make some ice—experiment with different containers or even try making ice outside in the winter—and then try this building experiment.

STUFF YOU NEED

Assorted plastic containers

Water

Freezer

Cookie sheet

Salt

THINGS YOU DO

❶ Make your building blocks. Gather a bunch of plastic containers and fill them with water. Stick them in the freezer (or outside on a cold day/night) and wait until they freeze.

❷ Build your ice towers! Empty all the containers and gather your blocks on a cookie sheet. Try stacking them. What happens?

❸ Now try sprinkling the ice blocks with salt. Start with a large piece for a base. Sprinkle salt on the surface and place another block on top of it. What happens? Add more and see how high you can build.

 WHAT'S GOING ON?

The ice is slippery. If you try to stack it, it will slide right off. When you add salt, it lowers the freezing point of the ice and a layer melts. Placing another block on top helps refreeze the water, and the two blocks stick together.

?   TAKE IT FURTHER

....

Pick up an ice cube with a piece of string. Sprinkle a pinch of salt on an ice cube and lay a piece of string across the salted area. Wait for a few minutes and see if you can pick up the string and the ice cube!

GEEK MAMA FUN FACT!

Did you know that you need salt to make ice cream? Water freezes at 32°F. But ice cream has more stuff in it than water—stuff like sugar and fat. To freeze extra stuff, you need to make things colder. It needs to be at least 27°F for ice cream to freeze. That means if you use just ice (which is 32°F) you're going to wait a loooooooong time and still never get ice cream. That's where salt comes in. It lowers the temperature. When you add salt to ice, you end up with a salty mixture that is actually colder than ice. Cold enough, in fact, to freeze your ice cream.

41

HOMEMADE EGG GEODES

What's cooler than smacking a rock in half and discovering brightly colored crystal geodes? Well, maybe making your eggshells into geodes! Here's a chemical reaction that grows crystals. Add some color and you get some amazing eggshell geodes.

STUFF YOU NEED

1 egg
Pushpin
Scissors
Small paintbrush
White glue
¾ cup alum (You can find alum in the spice section of the grocery store.)
Boiling water
2 small glass bowls or jars or plastic yogurt cups [CONT. →]

THINGS YOU DO

❶ Prepare the egg. Make a hole in one end of a raw egg with a pushpin. Make a few holes so you end up with an opening about the diameter of a chopstick. Make another hole on the other end. It doesn't have to be as big as the first hole. Place your lips on the end with the small hole and blow hard. The egg should trickle out the other side. Save it to eat (but use it within 24 hours).
❷ Use scissors to cut the shell in half lengthwise. Wash the inside of the shells. When they are dry, [CONT.→]

| Food coloring (I like to use neon food coloring for its vibrant colors.) | paint the inside with a layer of white glue and sprinkle it with a layer of alum. Let this dry for a couple of hours. |

Food coloring (I like to use neon food coloring for its vibrant colors.)
Paper towel

paint the inside with a layer of white glue and sprinkle it with a layer of alum. Let this dry for a couple of hours.

❸ Make the dye solution by adding ¾ cup of alum to 2 cups of boiling water. Stir until is it completely mixed. Divide this solution evenly into your 2 glass cups and add 2 to 3 drops of food coloring to each.

❹ When the solution is cool enough to touch, fully submerge each egg shell in a cup, cut side up. Let this sit overnight, for 12 to 15 hours.

❺ Carefully lift out your geodes. Place them on a paper towel to dry. Enjoy!

 WHAT'S GOING ON?

You are making a supersaturated solution by adding hot water to alum. As it cools, crystals form. Because you have the eggshells in the water with a thin layer of alum, the crystals will begin to form on these alum crystal "seeds," and all night the solution will form crystals. The color of the food coloring will be trapped in the crystals.

 TAKE IT FURTHER

Try using other containers. Does it work in small paper cups? Plastic eggs? Also try letting your eggs sit in the solution for longer. What happens?

GEEK MAMA FUN FACT!

Diamonds, rubies, and emeralds are all crystals that form when hot magma cools slowly. Salt is a crystal that forms when a salty liquid solution dries up.

₍4₂₎

CRYSTAL SNOWFLAKES

It doesn't have to be cold outside to make snowflakes. You can make your own snowflakes with pipe cleaners and a common laundry chemical.

STUFF YOU NEED

3 pipe cleaners
Scissors
String
Pencil
Wide-mouth jar
Water
Borax (Look for it on the laundry detergent aisle at the store.)

THINGS YOU DO

❶ Make the snowflake base. Twist 3 pipe cleaners together in the center to make a 6-pointed star. Use scissors to trim the ends of the pipe cleaners so they are all approximately the same length and can fit in the jar.

❷ Tie a piece of string to one end of the star. Connect the string to the next point by twisting it around the pipe cleaner. Continue around until you connect all the points together with the string, making a snowflake skeleton.

❸ Tie another piece of string to one of the pipe cleaner points, and tie the other end around the pencil. Place the snowflake in the jar with the pencil resting across the mouth of the jar. Make sure the snowflake hangs without touching any part of the jar. Take the snowflake out of the jar and set it aside.

❹ Make the snowflake solution. Use a teakettle or microwave to boil enough water to fill the jar. Measure how many cups of [CONT.→]

water are needed to fill the jar. For every cup of water placed in the jar, mix in 3 tablespoons of borax. This will make a saturated borax solution. Stir the borax solution with a spoon until as much of the borax dissolves as is possible.

❺ Create the snowflake. Hang your pipe cleaner snowflake in the jar so that it is completely covered in the solution. Let it sit overnight. Gently remove your now crystal-covered snowflake in the morning and hang it in an empty jar to dry.

 WHAT'S GOING ON?

When you mixed the borax with the water, you created a solution. The particles of borax mixed in between the water molecules. By mixing the borax into hot water, instead of room temperature or cold water, more borax can move between the water molecules. Hot water molecules move faster than cold water molecules and make more space in between them. The space gets filled with borax. When the solution cools, the water molecules come closer together and squeeze the borax out. As the borax begins to fall out, it starts to crystallize. It finds your pipe cleaner snowflake and sticks to it. More and more borax comes along and crystallizes on your flake and on top of other borax crystals until you pull your snowflake out of the water the next morning.

 TAKE IT FURTHER

+ To make colored snowflakes, use colored pipe cleaners and add 1 to 2 drops of food coloring to your snowflake solution.

+ To make your snowflakes glow in the dark, paint the pipe cleaner snowflake with glow-in-the-dark paint and let it dry completely before submerging it into the snowflake solution.
+ Tie a ribbon to one point of your snowflake to make a Christmas tree ornament!

GEEK MAMA FUN FACT!

Snowflakes are ice crystals that form high in the clouds. They always have six sides!

43

THERE'S A HOLE IN MY BOTTLE

OK, so you've got a plastic bottle full of water. Poke a hole in it and it drips, right? Wrong! Try this experiment and hold the bottle over your head. You still won't get wet. You've got atmospheric pressure and surface tension to thank for that. Try it!

STUFF YOU NEED

Pop bottle with cap
Water
Nail and hammer

THINGS YOU DO

❶ Fill your bottle right to the rim with water and tighten the cap. Tip it and look. No drips.

❷ Take the hammer and gently tap a nail through the bottle to make a hole in the plastic about ¾ of the way down. Pull out the nail. Is the bottle leaking?

❸ Hold the bottle over the sink and remove the cap. Now what happens?

 WHAT'S GOING ON?

There are a couple of science forces at work here: pressure and tension. Don't reach for the aspirin just yet; I'm talking about *atmospheric* pressure and *surface* tension. This dynamic duo makes it impossible for water to come out of the hole in the bottle when the cap is on. There is atmospheric pressure pushing in on us from *all* sides *all* the time. It never stops. When you have the cap on the bottle, pressure is pushing in on the bottle from all sides. It pushes down, up, and sideways. It pushes against the water in the little hole and prevents it from coming out.

But atmospheric pressure isn't the only thing that's stopping the water. There is this thing that happens when air and water meet called surface tension. It's like a kind of tough skin made up of water molecules that act as a bit of a seal. You can see surface tension when you look at a pond and see water-strider bugs walking on the surface. In the case of this experiment, surface tension and atmospheric pressure prevent the water from flowing. Unless . . . you take the cap off. That's a whole different ball of wax. NOW the atmospheric pressure can actually push past the mouth of the bottle and work on the water itself—pushing it down. This heavy push breaks the surface tension at the site of the hole and—voilà!—you have sprung a leak!

 TAKE IT FURTHER

+ Try Things You Do again. Fill up the bottle and put the cap on tight. Place the bottle in the sink. There should be no leaking. Open the cap and see the stream. Now hold a lit flashlight on the opposite side of the bottle as the hole and turn out the lights. What happens to the water? What happens to the light? Put your hand under the stream—what happens to the light now? Light will follow the stream of water.

+ Want more poking holes and not leaking? Try the Leakproof Baggie of Water on page 44.

44

MARSH-MALLOW LAUNCHER

I don't know about you but when we buy marshmallows we *always* end up with a huge bag that we never finish. They spend an eternity in the cupboard until the next time we look for a marshmallow. By that time, they are the texture of crunchy cat food. So we buy another bag. The cycle repeats. Until now! There is new life for old marshmallows! You can fling those marshmallows in the name of science! Harness potential and kinetic energy and shoot those little sweeties across the yard! I love science.

STUFF YOU NEED

Empty yogurt cup
Scissors
Balloon
Marshmallows (Mini marshmallows work best.)

THINGS YOU DO

❶ Cut off the bottom of the cup. This isn't easy if the container is sturdy. Be careful. Scissors can slip! *This step is for adults only.* Set the cup aside.
❷ Tie a knot in the balloon as if it were blown up. (But don't blow it up!) Cut the top ½ inch off the balloon. [CONT.→]

❸ Slide the open end of the balloon over the top part of the yogurt cup. Pull it up until the bottom is taut like a drum with the knot in the center.

❹ Load up the launcher. Place a marshmallow in the launcher. Pull the knot of the balloon while holding the cup tightly. Let it go. What happens? How can you launch a marshmallow farther? What happens with a handful of marshmallows?

 WHAT'S GOING ON?

Newton's Second Law of Motion and elastic potential energy are what's going on. But that sounds too complicated. Basically, it's a simple transfer of energy. You load the balloon with potential energy when you yank it back. Let go and the energy is transferred to the marshmallow that flings out of the shooter. The potential energy has transferred to kinetic energy. As for the kids—they'll get the idea through play: the more the balloon stretches, the more *kaPOW* and the higher flying the marshmallows will be. Put too many marshmallows in and they won't move much.

 TAKE IT FURTHER

What goes the farthest? Try flinging different things like raisins, peanuts, small candies, pennies. . . . Make a chart of things from light to heavy and measure how far each one goes. Record your findings. What did you discover?

(4 5)

SLIP-AND-SLIDE SCIENCE

Climb on up and whoosh! Slide on down. You're sliding, but you're experimenting with gravity and friction as well. Gravity is what pulls you down the slide. This experiment is all about exploring friction. Grab a couple of your favorite toys, such as blocks or cars, and toss them down the slide to see how friction affects them.

STUFF YOU NEED

Playground slide
Items to be tested (toys with wheels, wooden blocks, rocks, marbles, stuffed animals)

THINGS YOU DO

❶ Line up the things you want to test. Which one do you think will go the fastest? Which one will go the slowest?

❷ Test each item one at a time. Hold each item at the top of the slide and let go. Which one went the fastest? Which one was the slowest? Was it what you imagined? What qualities made some toys move faster than others?

Friction is a force that slows you down. It happens when two things rub against each other—like the seat of your pants and the slide. If there was no friction, anything that starts moving would never stop. If you slide your shoe across the floor, it's the friction that makes it stop. Things that roll may have less friction than things that slide. Things with a big flat surface may have more friction and will stop more quickly.

 TAKE IT FURTHER

+ Test your toys with waxed paper. Take one of the objects and rub the bottom of it with waxed paper. See how long it takes for it to reach the bottom of the slide. Try setting it on a square of waxed paper and see if that reduces friction.
+ Test your toys with water. Wet the slide and see what effect water has. Will the items move faster or slower? Does the amount of water matter? Time the trials with a stopwatch and make a chart.

46

MAKE YOUR OWN BOUNCY BALLS

Looking for a little bounce? Make it! Mix together a few simple household elements and create a rubbery substance that you can make into your own Super Ball.

STUFF YOU NEED

2 bowls
½ teaspoon borax
4 tablespoons corn-starch
2 stirring spoons
4 tablespoons warm water
1 teaspoon fluorescent or glow-in-the-dark paint
1 tablespoon white glue

THINGS YOU DO

❶ Make mixture #1. In a bowl, mix the borax and the cornstarch. Stir it together with a spoon so it's evenly mixed. Add the water and mix until you get a smooth paste. Set this bowl aside.

❷ Make mixture #2. In another bowl, add the paint and the glue. Mix it well.

❸ Combine the two mixtures. Let it sit for a bit and allow the ingredients to react.

❹ Stir the combined mixture. It will start as a sticky, hard mess, then it will turn a bit slimy. Keep mixing! Use your hands: roll it, squish it, shape it into a ball.

❺ Now bounce it! What happens if you let the ball sit? What happens if you put the ball in the refrigerator before bouncing? What surfaces are the best for bouncing?

❻ You can keep the goo in a zipper-lock sandwich bag in the fridge for a few days. The balls will lose their shape, but you can mold them back into shape quickly.

WHAT'S GOING ON?

By mixing the glue and the borax you have made a kind of plastic. Plastic is made up of long chains of molecules called *polymers*. The borax links up sections of the glue polymers and makes chains of molecules that stay together when you pick them up. It's a pretty gooey connection. In fact, without the cornstarch, this is a recipe for slime. The cornstarch helps to bind the molecules together so they hold their shape better.

 TAKE IT FURTHER

+ Tinker with your mixtures. What happens if you use more glue? What happens if you don't use cornstarch?
+ Want to make more goo? Check out Cornstarch Quicksand on page 84. What to play with polymers? Try the Leakproof Baggie of Water on page 44.

GEEK MAMA FUN FACT!

The original Super Balls got their amazing bounce ability from rubber compressed under thousands of pounds of pressure.

47

A GREAT PAPER AIRPLANE

Paper airplanes are wonderful. Transform a simple piece of paper into an object that flies. It's a perfect marriage of wonder, magic, and science. Here is a fail-safe paper airplane that always has wings.

STUFF YOU NEED

One 8½ x 11-inch sheet of printer paper

THINGS YOU DO

❶ Fold the paper in half the long way. (We call this the hot dog way.)
❷ Keeping it folded, fold the short edge of 1 side down to the first fold, making a 45° angle. Flip it over and do the same thing on the other side.
❸ Flip it over again. Fold the new fold you just created to the fold you made in step 1. Repeat for the other side.
❹ Flip it over and repeat the process. Fold down the new fold to the original fold. Repeat on the other side.
❺ Hold the center part of the paper airplane and open the wings out.
❻ Now throw!

Flight is a complicated thing. It involves a bunch of things, from the shape of the plane to the air. And flight also has to do with gravity, thrust, lift, and drag. For the kids, it's fun. Tweak the airplane and see how the flight changes, but if you really want to get technical, then here's your parent cheat sheet on flight. Any object needs to overcome four main forces in order to fly.

Drag: Hold your hand in front of your body with your palm so it is parallel to the ground. Swing your hand back and forth to the left and right of your body. Keep your arm straight. It should be parallel with the ground. Do you feel the air? Planes that push a lot of air, like your hand did, are said to have a lot of *drag*, or resistance, to moving through the air. Now turn your hand again, like you're slicing it through the air. You can still feel the air, but your hand is able to move through it more smoothly than when your hand was turned up at a right angle. If you want your plane to fly as far as possible, you want a plane with as little drag as possible.

Gravity: Everything is pulled down to Earth. Your plane has to be light so it can resist gravity.

Thrust: When you throw your plane into the air, you give it thrust.

Lift: When the air below the airplane wing is pushing up harder than the air above it is pushing down, that's called lift.

? : TAKE IT FURTHER

+ Hold a flying contest.
+ Throw the plane at least five times. How long is the longest flight?

+ Try making the plane out of smaller paper. Try making one out of larger paper.
+ What if you used a different kind of paper? Card stock? Newspaper?
+ Which planes fly the farthest?
+ Decorate your plane.

PAPER WHEEL
WIND WHIRLER

There's nothing like feeling the wind tousle your hair and kiss your face on a breezy day. Wind is a wonder. You can't see it, but with an old greeting card and a pencil, you can catch the wind and make it whirl.

STUFF YOU NEED

6 x 6-inch piece of card stock (An old greeting card works beautifully.)

Scissors

Pushpin

Map pins with a little bead at the end

2 sequins, small beads, or buttons (Make sure the pin can go through the hole.)

New unsharpened pencil with an eraser

THINGS YOU DO

❶ Take a square piece of card stock, fold the opposite corners together, and make a crease. Do this to the other side so you have a creased X in the middle of your paper. From each corner, cut along the crease until you get about ½ way to the middle.

❷ Gently bend the right top corner of each triangle into the center. Use a pushpin to press through all the flaps to make a hole. (Be more careful than I am. I always poke myself!) Keep the pushpin in place until you are ready to complete step 3. [CONT.→]

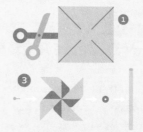

❸ Take your map pin. Place a sequin, small bead, or button on the pin, then take the pushpin out of the pinwheel paper and put the map pin through the hole (make sure you go through the hole in all four corners!), and then add another sequin, bead, or button. Press the pin into the end of the pencil eraser.

Put the map pin through a bead or a sequin, then the paper pinwheel, then another bead or sequin, and finally the eraser.

❹ Try blowing at the edges of the pinwheel to make sure it can spin. Loosen the pin a bit if the blades do not rotate well.

❺ Catch the wind. Go outside and see the breeze make your wheel whirl. What happens when you run with it? What happens when you turn it sideways facing the wind? What happens when you turn it to face toward the wind?

❶ WHAT'S GOING ON?

What is wind, anyway? It starts with the sun. The sun's rays beat down and heat up the Earth's surface. The Earth's surface, in turn, heats up the air above it. When molecules heat up, they move more and bump into each other more, making more space between the molecules. Because there is less space between the molecules, warm air is less dense than cooler air. So it rises up. Cooler air rushes in to fill its place. This is wind. Wind wants to blow straight. If you put the wind whirler in the wind, the wind can't keep blowing straight—it bumps up against the paper. The folded paper pockets catch the wind. The wind pushes the whole thing in a circle because it is pinned to a base in the center of the wheel.

❓ TAKE IT FURTHER

- + This is a great jumping-off point to talk about wind power. Look at windmills and discuss how the wind powers them.
- + Set up a fan and put it on the lowest setting. Hold the pinwheel in front of it. What happens? Turn up the speed of the fan. What happens?
- + Experiment with color. Draw red patches alternating with blue patches all the way around the wheel. See what colors you get when the wheel turns. Try yellow and blue, red and yellow.

(49)

SUN SHADOW CLOCK

We get a lot from the sun. The rays heat up our planet, help plants grow, and even give us power. You can cook with the sun, and you can even tell time with it. In fact, you can tell the time with nothing more than a stick in the ground!

STUFF YOU NEED

Stick

Soft ground or sand to
   push the stick into

Sunny day

Chalk

Pebbles

Watch

THINGS YOU DO

❶ Take your stick and stick it in the ground. Make sure you are in full sunlight and that your stick is straight up and down.

❷ Using the chalk, write the numbers 1 through 12 on your pebbles. These will be your time markers.

❸ Every hour, on the hour, place a pebble at the end of the stick's shadow. If it is 10:00, put the 10 pebble on the spot. At 11:00, put the 11 pebble on the spot where the stick's shadow ends. Continue to do this. Don't move a thing.

❹ The next day you won't need your watch. You can tell the time by looking at the shadows and the pebbles!

Why do the shadows change? To understand that, you have to know the basics about the sun and the Earth. I have to go through this checklist to remind myself why we have years, days, and seasons. It's a good refresher and a fun little game. Grab your kid, a flashlight, and a ball (or a globe).

First of all, the sun stays put and we rotate around it. Try this: Stand in one spot. Have your child stand about 3 feet away from you. You be the sun. Stay put. Have her move around you in a circle. She is the Earth. One rotation around the sun is a year.

But that's not all. We also spin on an axis. Try this: Stand in one spot. Have your child stand about 3 feet away from you. She'll be the sun first. Give her a flashlight and have her shine it at you. You not only move around her in a circle, you spin slowly as you do it. When you turn away, the flashlight doesn't hit your face. It's night for that side of the Earth. Turn toward her and you get a light in the face. Daytime! She stays put. See how she seems to move across your field of vision? Each full spin is one day. Because we are moving on our own axis the sun seems to move across the sky—actually it's not. We are spinning. Now have her be the Earth and you are the sun. You stay put, and she spins.

But wait, there's more! Our axis is tilted so we have seasons. Try this. Grab a ball—or a globe. A beach ball is good but any ball will do. Let your child still be the sun with the flashlight shining on the ball. Hold the ball straight up and down and keep your fingers on the poles. Now tilt the ball. Move around the sun—showing her that the North Pole is pointed at the sun for half of the time and then away for half of the time. This is why the shadows change slightly every single day.

 TAKE IT FURTHER

+ If you only have a sidewalk or a paved driveway nearby you, that can work, too. Have your child stand in one spot. Trace his or her feet with the chalk so the child can always be in the same spot. Do the same experiment. On the hour, every hour, wherever your son's or daughter's shadow ends, mark the sidewalk with chalk and state the time.

+ Try to observe the clock for a week or so. What do you notice about the shadows? Does it still keep time? Why not? The length of the shadows will change as the seasons change because the Earth is tilted and the angle of the sun's rays hitting the Earth changes all the time.

GEEK MAMA FUN FACT!

Before there were clocks, people relied on sundials to measure time by the position of the sun.

(50)

SWING AND SPIN

The allure of a playground never really goes away, does it? I still love swinging with my kids even though they're headed into teenager-hood. Of course, I can justify it because it's a learning ground for physics. But, really, it's so fun it will make your head spin—literally. Explore this simple activity with your kids—and make sure they're prepared to get dizzy. Did you ever wonder how ice skaters spin so fast? They start out with their arms out and then pull them in and zip around like a top. But why? Head to the swing set. This swing-and-spin experiment will leave you dizzy with science delight!

STUFF YOU NEED

Playground swing

THINGS YOU DO

❶ Trial #1: The spinner sits on the swing. The spinner twists the swing slowly around and around until the swing is wound up tightly. Count how many times you have to twist to get to this point so you can repeat it exactly. The spinner lets go. How long does it take for the swing to unwind?

❷ Trial #2: The spinner sits on the swing. The spinner twists the swing slowly around and around until the swing is wound up tightly. Make sure it is the same number of twists as before. This time the spinner stretches her arms and legs out. The spinner lets go. How long does it take for the swing to unwind?

❸ Trial #3: The spinner sits on the swing. The spinner twists [CONT.→]

the swing slowly around and around until the swing is wound up tightly. Make sure it is the same number of twists as before. This time the spinner pulls arms and legs in tightly. The spinner lets go. How long does it take for the swing to unwind? Which is the fastest?

 ## WHAT'S GOING ON?

When the spinner extends his or her arms and legs, some of the weight was placed out of the center. It made the spinner twist more slowly. When the spinner pulled in, all the weight was close to the center. When a spinning object is compact, it spins faster.

 ## TAKE IT FURTHER

+ Use a stopwatch and make time trials.
+ Next time your child sees skaters on television or in person, point out how they hold their arms when they spin. The closer their arms are to their bodies, the faster they can whirl.

51

SEED HUNT

My kids always found a challenge made anything more fun—a walk, a hike, or even a trip to the beach. Maybe it's the inner explorer in all of us that makes looking for something specific a huge adventure. This is a great thing to do when you're heading out for a walk. You can do it in the spring, summer, and fall, depending on what's in season. It's a fun challenge that can be used for a walk in the park, a hike in the woods, or a jaunt in the meadow—wherever there are plants.

STUFF YOU NEED

Notebook

Pencils

Old socks (Cotton or
    wool works well.)

Plant-filled place to walk

Magnifying glass

THINGS YOU DO

❶ Gather your notebook and pencils and state the challenge: you're going on a seed hunt.

❷ When you're ready to go for a walk, put the old socks on your hands. As you walk, brush your hands across the plants and see if you catch any hitchhikers. If you feel like it, put a sock on over your shoe and walk through the meadow. Did you catch any seeds?

❸ Once you have collected a few, look closely. What kinds of [CONT.→]

things do you see? Can you match the seeds to the plants? Notice the color, shape, and texture of the seeds.

❹ Record your findings. Make drawings of the seeds. Make rubbings of the leaves so you can identify which plants the seeds came from.

 WHAT'S GOING ON?

Every plant has seeds! But what are they and why are they? Seeds are packets of information and nutrition. The information is for growing a new plant, and the nutrition is for powering that growth. Seeds are amazing! Plants have developed a number of different ways to get their seeds around. If they didn't, then the seeds would sprout right under the parent plant and not get the sun they may need to thrive. Some seeds have sticky bits that catch on passing animals. (In fact it was this kind of seed that inspired Velcro!) Some seeds are surrounded by sweet fruit that is swallowed and pooped out miles away by bats, birds, bears, mice, and insects. Some have cool whirly bits that can catch the wind. Now go look.

 TAKE IT FURTHER

+ Try sprouting the seeds.
+ Press leaves and seeds and save them.
+ Make drawings of seeds and the plants they grow into.
+ Have your child tell the story of a single seed's adventure. Write it down and have your child illustrate it. Gather the pages and bind them together in a book.

52

NEWSPAPER TRIANGLE FORT

Who doesn't love making a fort? Blankets and pillows and the back of the couch can only take you so far. Channel your child's inner architect and use the power of the triangle to create an awesome fort with recycled newspaper.

STUFF YOU NEED

Newspaper
Tape

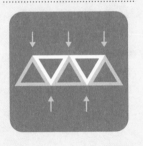

THINGS YOU DO

❶ Make newspaper tubes. Take two sheets of newspaper and place them flat, one on top of the other. Start at one corner and roll tight tight tight! The tighter the roll is, the stronger the tube will be. Roll to the end and tape it in place. Make a bunch of tubes. An ideal amount would be around 30.

❷ Make triangles with three tubes. Tape the corners so they are sturdy. Create as many triangles as you can with your newspaper rolls.

❸ Once you have your triangles, you can start to build. Place [CONT.→]

them edge to edge and tape them in place. You might want to tinker with the shapes before you tape them. The size of your structure will depend on how many triangles you made.

 WHAT'S GOING ON?

Engineers use triangles in a lot of structures. They are the strongest shape. Triangles don't squish. When you put a force on any part of a triangle it spreads out and pushes right into the angle. The angle won't change. They can withstand a lot of force. Other shapes are not as stable. Imagine putting force on a square; the angles can change easily, and it will collapse. But if you put a support diagonally inside a square to create two triangles, the structure is a lot stronger.

 TAKE IT FURTHER

+ Make sure you have walls and a roof. Use flat sheets of newspaper to create walls, or toss a blanket over the whole thing.
+ What does the fort look like? Are there separate rooms? What decorations do you have? Use your imagination!
+ If you want more building explorations, try Ice Cube Tower on page 104.

RESOURCES

Looking for more science fun? Try these books and websites.

BOOKS

Camp Out! The Ultimate Kids' Guide by Lynn Brunelle

Pop Bottle Science by Lynn Brunelle

Tinkerlab: A Hands-On Guide for Little Inventors by Rachelle Doorley

Naked Eggs and Flying Potatoes: Unforgettable Experiments That Make Science Fun by Steve Spangler

WEBSITES

Mama Gone Geek blog and more: www.lynnbrunelle.com

Steve Spangler Science: www.stevespanglerscience.com

The Exploratorium: www.exploratorium.edu

PBS Kids: http://pbskids.org

Bill Nye the Science Guy: http://billnye.com

ABOUT THE AUTHOR

A FOUR-TIME EMMY AWARD-WINNING WRITER for *Bill Nye, the Science Guy*, Lynn Brunelle has more than twenty-five years of experience writing for people of all ages, across all manner of media. Previously a classroom science, English, and art teacher for kids K-12, an editor, illustrator, and award-winning author of over forty-five titles (including *Pop Bottle Science, Camp Out! World Almanac Puzzler Decks*, and *Mama's Little Book of Tricks*), Lynn has created, developed, and written projects for Chronicle, Workman, National Geographic, Scholastic, Random House, Penguin, A&E, The Discovery Channel, Disney, ABC TV, NBC, NPR, The Annenburg Foundation, World Almanac, Cranium, and PBS. Her latest book, a memoir called *Mama Gone Geek*, was released in 2014 and won the Independent Publishing Award Gold Medal.

A regular contributor to New Day Northwest TV as a family science guru, Martha Stewart Radio as a family activity consultant, and NPR's *Science Friday*, she is the creator of the *Mama Gone Geek* blog and *Tabletop Science*—videos that make science fun and accessible.